Data Mining Mobile Devices

OTHER TITLES FROM AUERBACH PUBLICATIONS AND CRC PRESS

Data Mining Mobile Devices

Jesus Mena

CRC Press
Taylor & Francis Group
Boca Raton London New York

CRC Press is an imprint of the
Taylor & Francis Group, an **informa** business

AN AUERBACH BOOK

CRC Press
Taylor & Francis Group
6000 Broken Sound Parkway NW, Suite 300
Boca Raton, FL 33487-2742

First issued in paperback 2019

ISBN-13: 978-1-4665-5595-2 (hbk)
ISBN-13: 978-0-367-37989-6 (pbk)

This book contains information obtained from authentic and highly regarded sources. Reasonable efforts have been made to publish reliable data and information, but the author and publisher cannot assume responsibility for the validity of all materials or the consequences of their use. The authors and publishers have attempted to trace the copyright holders of all material reproduced in this publication and apologize to copyright holders if permission to publish in this form has not been obtained. If any copyright material has not been acknowledged please write and let us know so we may rectify in any future reprint.

Library of Congress Cataloging-in-Publication Data

Mena, Jesus.
    Data mining mobile devices / author, Jesus Mena.
    pages cm
    Includes bibliographical references and index.
    ISBN 978-1-4665-5595-2 (hardback)
    1. Data mining. 2. Mobile computing. 3. Web usage mining. I. Title.

QA76.9.D343M438 2013
006.3'12--dc23   2012050713

Visit the Taylor & Francis Web site at
http://www.taylorandfrancis.com

and the CRC Press Web site at
http://www.crcpress.com

To

Sergio "El Juez" Armijo, Jorge "El Mustachon" Bustamante,
Antonio "El Ochotres" Diaz, Eusebio "El Marcos" Gutierrez,
and Victor "El Romano" Munoz

Contents

Preface

This is a book about data mining mobiles in millions of peoples' pockets or purses — which represent incredibly powerful diaries of their life — *continuously* and *intimately* broadcasting where, how, when, and what products, content, games, news, movies, relationships, books, searches, services, interests, places, entertainment, etc., they want.

The structure of the book begins with the construction and leveraging of mobile sites, in Chapter 1, followed by the strategic use of mobile apps in Chapter 2. Both of these are important because they generate vital mobile data, which is the subject of Chapter 3.

Chapter 4 discusses mobile mobs, which can be differentiated as distinct marketplaces, that include Apple, Google, Facebook, Amazon, and Twitter. Finally, Chapter 5 discusses mobile analytics in detail via clustering, text, and classification AI (artificial intelligence) software and techniques.

Data mining, behavioral targeting, big data, business intelligence, Web analytics, and, most importantly, AI have been confined to stationary devices by enterprises, brands, and marketers. However, given the recent trends, this will soon change; the world is going mobile — and this is why this book was written. Here are a few numbers to consider:

- "Mobile marketing revenues will grow to $58 billion by 2014," Gartner <http://www.gartner.com/technology/home.jsp>
- "There will likely be more mobile Internet users in 2015 than PC users," IDC <http://idc.com/>
- "Global mobile marketing will grow by 30 percent to $1,047 billion in 2016," Ovum <http://ovum.com/section/home/>

Chapter 1

Mobile Sites

1.1 Why Mobile Sites?

Just as companies did, marketers and brands quickly realized a decade ago that they needed to create a website in order to have a presence and be relevant to consumers. History is repeating itself with the current mobile explosion; by 2013, more people will be using their mobiles than PCs (personal computers) to browse the Internet. The tipping point came in 2011; in that year, consumers spent more time on their mobiles than on their desktops or laptops—during that holiday season they searched, shopped and did price comparisons—via the devices they carried with them in stores and malls.

One critical feature about mobile sites is that they can generate a wealth of information for data mining mobile devices, such as the segmentation of different models and operating systems, or by their physical locations or interests. It is in this context that the importance of constructing and supporting a mobile site becomes paramount—as they generate important mobile data for modeling and predicting consumer behaviors and preferences. The mobile device *is* the consumer, and mobile sites are the means by which to use cookies and other mechanisms to track and segment their owners' behaviors.

In such an evolving environment, a traditional website designed to be viewed on a large stationary device comes across as woefully lacking on a tiny mobile device where a person may be walking down a street, maneuvering through a mall, or entering a store. New website designs are required when creating a mobile site—as those visitors are using small moving devices and new factors must be considered—such as *location-* and *interest-*based parameters. The features to include at mobile sites will differ among retailers, enterprises, and brands as consumers are no longer browsing with their mouse and keyboard but are instead swiping with their thumbs and tipping with their fingers.

One common dilemma when browsing via a mobile device is that traditionally designed websites take a relatively long time to load. The second common problem is that the print type on the pages is dismally small, requiring a lot of zooming to navigate the site. Additionally, certain style sheets and mechanisms from traditional websites such as Adobe® Flash®* are not supported by certain mobiles; this is definitely the case with all Apple®† devices. Mobile devices will overtake desktop use in less than two years, if not sooner; the number-one way people access local information on Google® or Facebook® is via mobile devices, with over 20 million users per month visiting just these two sites.

According to KISSmetrics (<http://www.kissmetrics.com>), 88 percent of all mobile users are more likely to buy from an auto dealer with a mobile site, 65 percent for auto parts stores, 62 percent for furniture, 61 percent for florists, and so on—the race is on. According to eMarketer (<http://emarketer.com>), almost 60 percent of all business-to-business (B2B) decision makers in the United States are on their mobile devices every day for email and Web research. They use mobile websites to make informed decisions—half of them are participating in social media, especially in industry forums in places like LinkedIn® or Twitter® on their mobiles—and another third are reading blog posts and listening to podcasts for information to support their everyday decision making.

Any enterprise or brand that does not have a mobile site is missing the attention of critical decision makers. Failure to construct one will result in the following: visitors will have to zoom in to read anything, probably multiple times, most likely there is too much text that no one will bother to read. Eye-catching calls-to-action will not be seen; the landing page forms will be too small to fill out on the mobile device; and finally, no one will take the time to browse a traditional website and navigate through layers of content. Mobile sites have to shout out, quickly and precisely; they need to get to the point instantly before the mobile browsers wander off to competitors.

1.2 Developing a Mobile Site

According to a recent survey from Compuware (<http://www.compuware.com>), 40 percent of mobile device users have turned to a competitor's site after a bad mobile experience. Yet currently, most companies and brands have not optimized

---

* "Adobe" and "the Adobe logo," if used, followed by other Adobe marks used in alphabetical order are either registered trademarks or trademarks of Adobe Systems Incorporated in the United States and/or other countries.
† Apple is a trademark of Apple Inc., registered in the United States and other countries.

their websites for mobile. A recent survey by Google Mobile Ads™ * of marketers and enterprises in 2011 found that only 21 percent have launched a mobile website. Google is eager to expand its online advertising empire further into the mobile ecosystem and has launched GoMo® † (<http://www.howtogomo.com/en/d/>), an initiative that aims to help businesses, marketers, and brands go mobile.

Google howtogomo.com (<http://www.howtogomo.com/en/d/#homepage>)is a clearinghouse of information on the topic of building a mobile site and includes features that allow marketers, brands, and enterprises to see how their site looks on mobile devices. For those looking for a quick fix, Google provides a list of companies that specialize in constructing mobile sites—visitors can specify what they want to spend, and some can get a site up and running for as little as $100 a year.

Mobile devices introduce a new layer of complexity that can be difficult for traditional websites to accommodate because they require *cross-platform* functionality and diminutive displays; at a minimum, the mobile sites should support the native Android™‡ and Apple browsers. Mobile-optimized sites need to simplify content and make it highly readable. They also can take advantage of audio-video content instead of long pages of text—which for mobile devices is like viewing a TV ad—as opposed to browsing a static website. Mobile sites need to draw in consumers in a fun and intuitive way, keeping in mind that they are also a new ad channel for delivering very important lead generation content to millions of consumers.

Marketers, enterprises, and brands need to consider how the mobile site structure will be organized and indexed on the Internet. There are three basic domains options for a mobile site: the first is constructing an independent site just for mobiles, while the second option is creating a subdomain or subdirectory for the mobile site, and finally creating a hybrid of both the traditional and mobile sites.

A separated mobile site is hosted on a different domain than the main website and works in a completely separate way from the traditional main website. This option increases the cost of creating, maintaining, and updating a separate mobile site, which tends to be higher than the other structure options. In addition, the separate new mobile site will not benefit from the traffic, links, or rankings of a traditional website. The new, separate mobile site must be positioned by the mobile search engines and must compete with other mobile sites that are already well positioned in these engines.

The second option of creating a subdomain or a subsection for a mobile website represented in the main URL (universal resource locator) may be done in the following way: *name-subdominio.dominio.com*. Subdirectories are sometimes called folders or subfolders and are represented differently in the URL. One of the biggest advantages of this option is the ability to customize the content of a main website

---

* Google Mobile Ads is a trademark of Google Inc., registered in the United States and other countries.
† GoMo is a trademark of Google Inc., registered in the United States and other countries.
‡ Android is a trademark of Google Inc., registered in the United States and other countries.

to the needs and requirements of the mobile Web. This usually means eliminating unnecessary components and controls, and most importantly the elimination of multimedia content that slows down the loading time for mobiles. Another option is the simplification of the mobile site design into a single column with different headers correctly labeled.

The hybrid option is based on the use of multiple cascading style sheets (CSSs) to adapt the presentation and the contents of the traditional website to mobile devices. The CSSs allow the developer of the site to determine how the content of a webpage is displayed on a mobile device. The site developer will need to insert a link to the mobile CSS for each page after the link to the traditional style sheet in order for the mobile browsers to automatically detect if there is a style sheet adapted for mobile devices.

Today, most of the modern mobile devices support Wi-Fi Protected Access II (WPA2), which is a security protocol and security certification program developed by the Wi-Fi Alliance (<http://wi-fi.org/>). WAP2 technology also allows the viewing of images, XHTML, and CSSs; in addition, WAP2 offers monochrome or high-resolution displays supporting CSS and Java™*. The use of this technique is very simple, especially if the code for the host that controls the styles is in a separate file server. This design allows for site updates to be done automatically and quickly. These updates will be taken by the mobile browser to display the page in an optimal way. This minimizes the load time of the mobile site because it is only necessary to load the style when the developer accesses the site but is not required for subsequent accesses to other pages.

Keynote Systems (<http://keynote.com>) released some mobile site performance metrics that indicate important industry best practices. The mobile metric company said the top mobile site performers stood by a couple of key mantras in managing and developing their presences. The first and most important best practice is that *less is more*. It is as simple as this: the fewer number of objects on a mobile home page, the less time it takes to load. On a small mobile screen, the user experience is about *clarity, usability,* and *speed*. Images and objects look great on a desktop, but on a mobile they just take a long time to load. The best practice for mobile sites is to cut down their loads by reducing the number of heavy images and using clear, directive text for their mobile homepages.

Here are a few things that developers, companies, marketers, and brands should not include in their mobile sites:

1. Avoid multiple forms; keeping forms to one page allows mobile visitors to see exactly where they are in the process of filling out the form.
2. Avoid irrelevant functions; instead, concentrate on quick functionality, such as click-to-call or click-to-text.

---

* Java is a trademark of Sun Microsystems, Inc.

3. Avoid all pop-up ads; this predicament can be caused when a company chooses to sync its traditional website to its mobile site.

It is important that all of these mobile-specific options be simple and efficient.

As companies, brands, and marketers extend their businesses through mobile websites and apps, they should be careful to use mobile-specific strategies that put the end-user experience first. All obstacles should be removed in order to allow end users to easily engage with a company's mobile site and apps. Allowing visitors to engage with key business processes at mobile sites can add significant business value to brands, enterprises, and marketers. Traditionally, this high level of mobile engagement has required the creation of native mobile apps, a time-consuming and costly building process that results in siloed data and inflexibility.

However, firms such as UR Mobile (<http://urmobile.com/>) are offering an attractive alternative to native business apps, and that is mobile site apps. Through its Accelerate™ Platform, UR Mobile allows companies to quickly integrate a mobile Web app that can be embedded in their mobile site. With mobile Web apps, there is nothing for mobile end users to download or manage, and the Web app solutions can be easily and efficiently updated and edited at any time by the company, brand, or marketer without having to go through a re-release or app store approval process. This option allows for two data aggregators to be created and evolve—the mobile site and the mobile app—both with one strategic stroke.

1.3 Mobile Biz

The number of mobiles is rapidly advancing; in 2011 alone, the tablet market experienced explosive growth, thanks to the Apple iPad® and Kindle Fire™*. The acceleration of these mobiles will significantly increase many aspects of business, including content consumption, customer touch points, and mobile commerce. Today, millions of apps, publications, sites, social networks, and other digital media content are readily available with the swipe or a tap of a finger. The question for the marketer, developer, and brand is: How will they leverage these mobile opportunities to foster more productive relationships with customers and prospects?

While mobiles support an "always on, always connected" way of life, they help people navigate the day in ways they never could have imagined. Beyond mobile sites and apps, consumers are now searching the Internet—Google sees 4 billion local searches for products and services each month with 61 percent of those resulting in an actual purchase. This is important for small retailers because 55 percent of consumers report using their mobiles to buy a local service or product.

---

* Amazon, Kindle, Kindle Fire, the Amazon Kindle logo, and the Kindle Fire logo are trademarks of Amazon.com, Inc. or its affiliates.

Every business is a little different, but this much is becoming clear: at some point in that sales cycle, people are turning to the Internet for more information. So it is mission critical for a marketer, company, or brand to understand the basics of local mobile search optimization and put these skills to work. This starts with optimizing their mobile site at all search engines, with ample maps, directions, discounts, and calls to contact touch points with one-click functionality.

Enterprises, retailers, developers, and brands also need to start considering the acceptance of mobile payments and this should be planned as an option and development for the mobile site. Expect mobile payments to increase and for mobile site visitors to begin asking for that mobile option—thanks in part to continued pressure from major players like Amex™*, PayPal™ †, VISA® ‡, Intuit®, and Google—plus, start-ups like Square® § and Dwolla® ¶, which are accelerating this payment option.

People like to watch video—in fact, YouTube™ ** is the second-largest search engine, behind only the behemoth, Google. People do not mind watching video ads on their mobiles; as Hulu™ ††, Netflix™ ‡‡, and YouTube apps continue to provide quality mobile content at a very low cost, consumers will become even more used to watching video on-the-go. There is an entire sector of the population that turns to YouTube to search and be entertained. Marketers, brands, and developers need to begin to develop a content strategy for social sites such as Facebook and Twitter and search sites such as Google and Bing® with equal attention to YouTube, which is ideal for advertising to mobiles on-the-move.

Here are some other issues to keep in mind when building a mobile site for getting down to business. Do not just scale down a traditional website and try to squeeze as much content into the little screen; instead, use the mobile development process to prune original offerings down to the most essential for mobiles. Put content over navigation; nobody cares about an enterprise's organization chart; instead, use screen real estate to display actual content, not just hierarchies of links. Consider what your visitors want first and then make sure you can deliver it: don't start from what you have; start from what they need.

Design the site to make the mobile experience fast; people use mobiles in between other tasks and nobody wants to wait while they are waiting. Check what visitors regularly do while at the mobile site, especially when changes are made. Do not assume to know how visitors will use the mobile site; rather, rely on mobile site analytics to measure what visitors do. The mobile landscape is constantly evolving

---

* Amex is a registered trademark of the American Express Company.
† PayPal is a registered trademark of PayPal, Inc.
‡ VISA is a registered trademark of Visa Inc. in the United States and other countries.
§ Square is a registered trademark of Square Inc.
¶ Dwolla is a registered trademark of Dwolla Corporation.
** YouTube is a registered trademark of Google Inc.
†† Hulu is a registered trademark of Hulu, LLC.
‡‡ Netflix is a registered trademark of Netflix, Inc. (http://www.netflix.com).

in terms of devices and even operating systems, so use HTML5 and other Web standards to future-proof the site development.

Mobile devices are becoming more capable of rendering pages on the front end, so use this ability to limit server calls and provide for each device's unique implementations. Finally, to make more screen area available for content is to tuck navigational options behind clearly labeled buttons that allow users to explore options and dive deeper. The developer can make navigation disappear altogether by setting up preferences for mobile alerts; visitors will get the content faster and the developer will learn about what they want.

1.4 Put Your Brand in Their Pocket

As more consumers engage through mobile devices, marketers are looking for ways to communicate with them—whether through apps or mobile sites where visitors are looking for content and information. For example, AvatarLabs (<http://www.avatarlabs.com>) recently created a series of mobile sites to help train retail sales associates at Best Buy® and AT&T™ so they could help consumers in their stores. Through the sites, employees accessed real-time information about products and resources to help clients on the sales floor.

Meanwhile, a new release from iBuildApp (<http://www.ibuildapp.com>) could help build buzz about a brand's mobile offering. Their new widget previews a brand's app on a mobile site, so potential users can "see" how the app will look and the types of information offered. This could help their decision making in downloading the app—more informed decisions mean a higher likelihood that those apps will be used rather than downloaded and deleted or ignored —it is a great way to reach customers and visitors, and get a brand in their pocket.

Finally, in an effort to help all those users with fitness resolutions, a new augmented reality program from Hipcricket®* (<http://www.hipcricket.com>) offers athletes training tips, sports, and information. Through a mobile ad campaign, consumers use their mobiles to access an interactive training card. That card "comes to life," offering the user training tips and athlete and sports information — the mobile ad campaign incorporated a number of channels to provide readers with a truly interactive and engaging experience. This is another example of how to make the mobile experience truly engaging for and rewarding to visitors and customers.

mytaGGle (<http://www.mytaggle.com>) offers a completely free Web application for creating a mobile site and apps for marketers and developers. The company provides businesses and individuals with the opportunity to swiftly create a professional app and site optimally suited for visitors using mobiles. Already half of all Internet users use mobiles—and this percentage will only rise further; at the same time, it seems that only a small fraction of all traditional websites can be

---

* Hipcricket is a wholly owned subsidiary of Augme Technologies, Inc.

viewed properly using mobile devices. When using the mytaGGle Web app, it soon becomes apparent how complete the package actually is.

Apart from the wide choice of templates and icons currently available, it offers to developers the option to design their own icons and to define the mobile site layout. Developers and marketers can easily and neatly position photographs and YouTube films on their mobile sites with links to social media, a Really Simple Syndication (RSS) page, which is a family of Web feed formats, or a Google page. They have even developed a unique link with Facebook that enables the developers or marketers to integrate their entire app site into their brand's Facebook profile. mytaGGle is a completely free service, and it only takes developers a few minutes to create an attractive app site for mobile devices.

Another firm, myhosting.com®, has launched its Mobile Website Builder platform, goMobi, enabling developers to build a professional mobile site both quickly and easily. The goMobi Mobile Website Builder platform includes a number of features through a WYSIWYG (What You See Is What You Get) interface that allows customers to easily create a mobile version of their traditional website by adding options like Products, Contact Forms, Find Us, Reviews, Image Gallery, and many more features. The goMobi Setup assistant can also automatically generate a mobile site based on an existing traditional website, to help visitors and developers get started quickly.

With mobile Web browsing becoming a more common way for users to access the Internet, the time to create a mobile-specific site has never been better. The goMobi Mobile Website Builder and Setup Assistant allow developers to create a site for smaller screens, enabling marketers to more easily engage mobile website visitors with a customized browsing experience. With search engines such as Google indicating that mobile is an upcoming trend that webmasters should be aware of, the time to jump on the bandwagon is now. Mobile users interact with search results and sites differently than desktop users.

Another mobile marketing company, ConnectMe QR (<http://www.connect-meqr.com>), offers its mobile website and QR Code package subscription service. This service also makes it possible for individuals and small businesses to have a mobile marketing site that they manage themselves, giving them the opportunity to compete in the mobile marketing arena often dominated by big business. The ConnectMe QR mCard™ comes with a unique Quick Response (QR) Code and URL, allowing developers and marketers the flexibility of attracting both mobile and online visitors.

Each ConnectMeQR user receives an mCard mobile site that is hosted by ConnectMe QR and has the ability to customize and control content by adding banner images and editing icons and text. Also included are nine different links developers can tailor for their site, such as phone, site, email, map, and social media. A unique ConnectMeQR Code can be printed on business cards and marketing materials, including access to their mCard mobile site 24/7 with edits and pricing, contact information, and marketing campaign changes taking effect immediately.

Here are some tips for getting your brand in their pocket or purse. Plan your mobile marketing strategy based on how people use their mobile devices. Do some market research, or perform a survey of your visitors' mobile habits: what are they searching for, or what are their desires, interests, and locations. The developer should evaluate how to target market using mobile technology and how mobile marketing will uniquely allow marketers to reach their goals. For example, a very common use is to locate a restaurant or other business that is close to where the mobile is located at the time, utilizing the GPS (Global Positioning System) or Wi-Fi triangulation functionality of the device.

1.5 Mobile Search Engine Optimization (SEO)

If you want them to choose your location, make sure that your online directory listings are up-to-date, accurate, and are search engine optimized (SEO). If you do not have online directory listings, get them. For example, if you own a restaurant, you will want to be on online directories such as Yelp™ (<http://www.yelp.com>), Google Places™ (<http://google.com/places>), Zagat.com (<http://www.zagat. com>), Citysearch.com, Opentable.com, or Insiderpages.com. Once you establish whether or not mobile marketing is appropriate for your marketing campaign, you should integrate your mobile marketing with your mass media and outdoor marketing by placing banners and ads that reference your mobile-friendly site in places where people might have idle time with their mobile device, such as at bus stops, in subway stations, cafes, or on billboards.

One of the easiest ways to begin mobile marketing is to ensure that your mobile site is locatable via mobile search engines and usable on mobiles. First, you need a site that is optimized for mobile devices. The elements of a dynamic traditional website are different from those of a mobile site. The trick is to make it attractive and functional to both kinds of visitors, but keep in mind that increasingly these will be mobiles. Be sure to test it on all the major platforms to ensure that it works as intended; this includes the two dominant devices: Android and Apple. Microformats are a newer class of open-data formats that can be used to adapt existing data so that it is more flexible for use across multiple platforms.

With mobiles becoming increasingly popular, it is becoming more important to take into account mobile-specific SEO when it comes to designing a mobile site. It is wrong to assume that mobile SEO is the same as SEO used in stationary computers; this is definitely not the case. To begin with, the content and method of presentation differ greatly between traditional websites designed for stationary devices and those designed for mobiles. The main difference is in the screen size, but the environments between these stationary and mobile devices are also different—mobile Web browsing takes place in one of two situations: "on the move" or "sitting down."

The search engines that have been built specifically for use on mobile devices have also been built differently than those created for use on stationary devices. The user's location, types of mobile device being used, and the content format are more critical to SEO than keywords in mobile search engines. This, however, is not to say that traditional SEO techniques should not be applied to mobile websites. Include keywords in title tags, headings, and content, as well as the SEO principles used in traditional websites.

There are some areas where mobile website SEO differs from that of traditional websites, and those areas include that the most important content needs to be placed in the top part of the webpage to ensure search engine spiders, and users, can find it easily. The CSS must be used for the layout in mobile websites, as tables will not render well at all, valid XHTML (eXtensible HyperText Markup Language) coding must be used, even though WML is the mobile-specific language; XHTML is more readable. Finally, it is recommended that the maximum size of any given webpage for a mobile device not exceed 20 kilobytes.

It is also important to ensure that that a mobile site focuses on *localized* SEO strategies. This is for a number of reasons: location is highly important in the ranking of sites in a mobile search. Location-based SEO on a mobile site will ensure better results—mobiles are most often used by people on-the-move who want to find something locally in a hurry. Localized SEO will ensure that potential visitors find the site based on location, the type of mobile being used to browse, and the format of the content. Mobile SEO expands the attributes from mere keyword tags to include the mobile's (1) location, (2) type, and (3) content.

1.6 Mobile Site Requirements

Developers and marketers need to remember that mobile visitors and all major search engines are using the Internet in different ways and for different purposes— mobiles are not PCs. Mobile users are accessing the Web using a small keypad, predominantly with one hand—and perhaps with a pointing device, such as a finger or a stylus—and oftentimes while balancing on one leg at the back of an overcrowded transit vehicle during rush hour. This tends to make a difference in their patience thresholds, and also makes SEO, usability, and search critical to a mobile site's marketing effort.

The objective of a mobile site design is to make it a simple and elegant user experience!

Keep in mind that mobile visitors may be viewing the mobile site through a different screen via a different browser on their mobiles. This may be a fully featured combination such as that found on Apple devices, or it may be a stripped affair of cheaper mobiles. This makes a critical difference in how much content mobile visitors are able to consume in one go and how much of it can be rendered by the developer. This means that mobile Web production efforts are more complex as debugging and testing user experience become much more than a quick sanity check in Microsoft® Internet Explorer® (IE), Google Chrome™, Apple Safari®, and Mozilla Firefox®.

As previously mentioned, mobile sites are accessed by users on the move; being on the move means users who are likely to be performing several tasks, such as searching for a particular restaurant or a particular piece of information (e.g., a flight departure time). Mobile site visitors may also be stationary, where they are likely to be engaged in a more private, immersive browsing experience, such as watching a movie, socializing with friends, or catching up on the news headlines. Just as mobile Web users are different, most mobile-specific search engines, such as Google Mobile™ (<http://www.google.com/mobile>), Jumptap (<http://ww.jumptap.com>), Medio™ (<http://www.mediosystems.com>), and Taptu® (<http://www.taptu.com>), are also built in different ways.

Mobile SEO exists on a different plane: keywords are still critical, as search engines will always operate on that basis, but newer dimensions such as location, device types, and content formats are more critical to and indicative of the mobile Web experience. Table 1.1 illustrates some of these differences.

Table 1.1   Mobile versus Desktop Differences

Search Dimension	Mobile	Stationary	Mobile SEO Challenges
Keywords	Limited	Many	Ensure relevant content is based on sketchy user input
Locations and Interests	Critical	Not Critical	Simplified presentation of results content in relation to a mobile's immediate location and needs
Browsers	Varied	Standardized	The challenge is the presentation of content in an accessible way and assurance of a good user experience, regardless of multiple mobile devices and browsers
Content	Poor	Good	Delivering a high quality mobile experience from poorly formatted raw mobile site source material
Content Formats	Specific	Generic	Using device information to help improve results by serving relevant content formats

With the above in mind, here is how search engines are beginning to adapt to mobiles searches with limited and fewer keywords. According to Google research, the average query search on Google Mobile is fifteen characters long—and this takes roughly thirty key presses and approximately forty seconds to enter; this means that search engines do not have a lot to work with when tasked with providing the user with an experience that equates to the quality of a desktop search. However, search engines are adapting to mobiles because that is where the traffic is heading.

One way in which Google and others are compensating for this lack of keyword action is by providing what is known as "predictive search," or predictive phrase query suggestions based on text mining and clustering technology. This helps users complete their queries more easily and also helps them deliver more relevant mobile search results. For example, a search for "pizza restaurant, San Francisco" on a mobile can result in triggering a variety of predictive suggestions that attempt to complete the search query with a variety of options such as: "pizza restaurant," "pizza north beach," "pizza take out," and so on.

This text mining application is a way that search engines assist mobile users in conducting a faster search and guaranteeing the results of relevant content based on location and interests. For the mobile site developer and marketer, this new functionality presents a new SEO opportunity, because by properly designing their site with the right metatags and the set of keywords in critical categories in the most common "predictive search phrases" that are specifically related to their business, brand, company, or location.

Because a large majority of mobile searches are locations, interests, and task specific, search engines are beginning to present their content in new ways to make results more accessible to mobiles. Using Google Mobile as the example—the results page is normally restricted to a list of five sites—their research has shown that mobile users do not tend to browse deeper than two pages. So, much like the new predictive search functionality previously mentioned, what today's search engines are doing is making some educated guesses on the user's behalf to try and get them to the results that Google thinks they want and as quickly as possible—remember that these searchers could be walking down the street!

In terms of the presentation of search results, Google—using GPS and Wi-Fi triangulation—recognizes that when a mobile is searching for a pizza restaurant in the Bay Area, it gives a heavier bias to those mobile sites that it thinks satisfy these location- and interest-based parameters. It gives them "featured presentation" treatment to these top-ranking sites, as they are the most relevant to the mobile user and their Google clients. Additionally, in order to compensate for small screen sizes, search engines are dividing the presentation of their results content into new location- and interest-based design layouts.

Additionally, from the search engine's point of view, this makes the mobile environment a challenge, particularly if, like Google, they are trying to recreate the slickness of their desktop experience on the mobile platform. To compensate for this relative anarchy, these engines are using a couple of techniques to make the problems disappear: site transcoding and user agent detection. Google Mobile, AOL®, Windows Live™, and others use transcoding software in order to give their users a more uniform user experience. In practice, this means that they have decided to impose mobile website presentation standards of their own; and if your site does not conform to them, then they will take your content and repurpose it to the design, layout, and format that they feel is best suited to the user's mobile. A transcoded version of a mobile site means that webpages are hosted temporarily on the search engine's servers and domain, rather than on the developer's site, with the URLs and links also transcoded.

User agent detection is another form of transcoding; it takes a mobile site's content and, if necessary, re-purposes it in the name of providing a more uniform browsing experience for various mobiles. The implication of all this transcoding work is that those mobile sites that avoid it by conforming to a more standardized means of mobile presentation will probably fare better when it comes to search engine and ranking.

Mobile users conduct their searches using a disproportionately high volume of brand names and, more obviously, location- and interest-based phrase categories. As already mentioned, the mobile Web is a different beast, used for different purposes, where people are searching for different sorts of things using a different sort of language and techniques. For the mobile site developer and marketer, this has a couple of very important ramifications: new search relevance is now determined by more *immediate* dimensions, such as location or vertical products, services, or brand suitability. For this reason, it is important for mobile site developers to understand the mobile-specific search phrases and interests that mobile users are using in order to optimize their content around those key terms.

This is important because consumers more often turn to a company or brand mobile site than to an app for shopping, according to a report conducted by Nielsen that tracked the shopping habits of thousands of iOS® and Android™ mobiles. Nielsen.com found that retail mobile websites are more popular than retail apps, and Amazon's is the most popular retail mobile website of all. Behavior split slightly along gender lines; the survey found that of those who did try a retailer's app, men were more likely than women to do so.

Target® and Walmart® skew female when it comes to their mobile websites, while Best Buy skews male, and Amazon® and eBay® appeal to both genders—retailers need to think that their businesses can potentially include mobile, online, and bricks-and-mortar stores. Winning with shoppers requires a consistent experience across mobile sites that reinforces the values of a retail brand, whether it be price, service, reviews, selection, style, or other key attributes.

At the same time, much like the early Internet desktop users, consumers are happy to take some guidance in overcoming their navigational challenges—lack of time, small form factors, and screen size restraints. We have already looked at "predictive search queries," but in terms of navigational aids, it is worth noting that search is definitely not the only component. Search may not even be the number-one activity on the list when a user is trying to locate stuff on the mobile Web; their primary interface is likely to be their operators' portal or a bunch of prepackaged vertical directories. To this end, this type of Web browsing service is currently at least as important as Google, if not more.

Ensure that the mobile site can be crawled at the code level—use the correct headers, do not block Internet Protocol (IP) ranges unnecessarily, use the correct robots.text file instructions—and ensure that all of the pages to be indexed are situated in the public domain. Submit the mobile sitemap to Google, Bing, and Yahoo!® in order to help them discover it and give them a head start when it comes to crawling and indexing. The developer should ensure that the navigation scheme is easy to crawl by coding it cleanly and ensuring that all the key content sits some-where within easy reach of the top-level pages. Ensure that the content contains a sensible level of outbound links that lead to other complementary and preferably related mobile site pages.

This is a basic approach to traditional search engine optimization (SEO) that appears to be overlooked when it comes to the mobile sites. It is easily explained by the relative value of on-screen real estate—the desktop Web affords more screen space in which to present outbound links, whereas the mobile experience puts screen space at a far higher premium. Submit the mobile site to DMOZ (<http://www.dmoz.org>), the Open Directory Project that is maintained on an open-source basis by human editors and used as a seeding index for many mainstream search engines; if the submission is accepted, it will improve the chances of mobile search engines picking up the domain and starting to crawl the site. Encourage other, related mobile sites to link to the site—using markup that is helpful to the overall keyword marketing strategy. The theory of page rank will continue to flourish on the mobile sites as far as search engines are concerned—as illustrated via the pro-prietary Google PageRank™ algorithm (<http://www.google.com/technology/>).

Ensure that the content layout is suitably simple for a mobile audience. Do not use Flash, Ajax, or other presentation methods that may make sense on a desktop, because they render the mobile experience cumbersome and should be avoided. Mobile crawlers will largely follow the browsing patterns and experiences of humans—burying key content in inaccessible layers of mobile pages can create a struggle for search engine crawlers. Do not make anything or anyone work too hard to access the mobile site content; different mobiles and different browsers will splice the mobile content in different ways.

Some search engines will decide to transcode it; the best way to make the content accessible is to keep it simple—make page titles, subheaders, content extracts, images, and body copy suitably concise, pithy, and readable to mobile crawlers. Conform to the new W3C mobileOK (<http://validator.w3.org/mobile>); these guidelines provide all the code-level instructions needed to make a content mobile site ready. They cover everything from the creation of mobile-friendly style sheets (CSSs) to the correct rendering of tricky content elements such as tables and image maps.

For further information, see <http://www.w3.org/TR/mobileOK-basic10-tests/>. While these guidelines were created to help mobile webmasters ensure that their sites would be accessible to mobile devices, they are also a critical part of an SEO strategy because they represent the standards around which most major search engines build their indexing algorithms. A mobile site is more likely to be recognized, indexed, and rated as a mobile site if the code is mobileOK compliant.

Use compliant markup language to ensure that the widest range of mobiles can access, read, and render the site content. This means WML (Website Meta Language; <http://the wml.org>), WAP 1.0 (Wireless Application Protocol; <http://www.wapguru.in/wap-technologies.php>), xHTML "Mobile Profile" (<http://www.w3.org/TR/SVGMobile>) or WAP 2.0 (<http://www.wapforum.org/what/WAPWhite/_Paper1.pdf>), and/or cHTML (<http://www.w3.org/TR/1998/NOTE-compactHTML-19980209>); good use of standards-compliant code will again ensure that search crawlers can easily find and index the mobile site and thus make a better candidate for inclusion in search results. Create the site content targeting mobiles with moving audiences in mind. This means paying more attention to shorter form factors for key important SEO content fragments such as URLs, page titles, and metadata—all of which will be re-used in search results pages and thus need to be suitably keyword-relevant to the search query in question—but also concise enough to be rendered and read on a mobile screen.

Many mobile search engines will help to achieve this presentation by removing standard elements in URL strings such as the "http://" to maximize the use of mobile-targeted page content. All major mobile search engines are now beginning to build in new mobile dimensions to their indexing methods and the presentation of their results, such as location indicators, content formats such as ringtones, and anything else that identifies the site as mobile-relevant or mobile friendly. One key way in which the developer can identify their site content is to use new "micro format" and "semantic" markup standards such as "hCards" (see <http://microformats.org/wiki/hcard>) that enable a mobile webpage content to be picked out—for example, the information within a "Call Us Now" button on the mobile site—as an hCard; this makes it possible for the mobile browser to re-purpose the source data for dialing up a phone number; for further information, see <http://microformats.org>.

1.7 A Mobile Site Checklist

- Use 100 percent valid XHTML 1.0 code. Many optimizers on the traditional WWW (WorldWide Web) do not consider using valid code as a best practice. Mobile search engines, however, may have more trouble digesting invalid code, so validate that the mobile search engines will not have any trouble with the site.
- Follow accessibility best practices. These will ensure that the content is accessible to anyone, regardless of their platform; this includes mobile users'

browsers and *mobile search engines.* The W3C Web Accessibility Initiative (<http://www.w3.org/WAI>) is a good place to turn to for the latest information on accessibility, and Dive Into Accessibility (<http://diveintoaccessibility. org>) is also a useful tutorial.

■ Follow traditional on-site search engine optimization (SEO) best practices, with major keywords in the title tag, H1's and body text, keyword-rich anchor text for internal links, etc.

■ Get spidered and indexed by mobile search engines; submit the mobile site to all the major mobile search engines for quick mapping: Google Sitemaps™ (BETA) and Yahoo! Submit Your Mobile Site. It is also a good idea to ensure that each of the mobile webpages has at least one incoming link.

■ If you are using a content management system such as WordPress® (<http:// wordpress.org>) building a mobile version of a site can be as easy as installing a plugin; simply search for "mobile" and pick a plug-in that best suits your needs.

■ There are numerous services available to create an app for mobiles or to make a mobile site; depending on your budget, these apps could be extremely intricate or as simple as a landing page with basic information.

■ If you are not using a content management system, you will need to enter some code onto your existing website that tells it to behave differently if it is accessed by mobiles.

■ Be sure to check for compatibility—at the very least, across Apple and Android devices—because there are many more mobiles on the market.

■ Consider using goMobi™ (<http://gomobi/>), a service designed for small and medium businesses to easily create a user-friendly mobile site that works on all major mobiles.

■ Studies have found that apps are mainly used for navigation and information gathering by the user, while mobile sites are preferred for entertainment and search queries; depending on the content and industry, this should be part of a brand, retailer, and enterprise strategy decision.

■ A report from Jumptap Jumptap (<http://www.emarketer.com/Article. aspx?cR=1008825>) found that businesses looking to boost their SEO should focus on their mobile sites; the study found that users conducting searches would rather access a mobile site than download an app.

■ Consideration should be given to formulating a mobile matrix strategy (Figure 1.1).

■ Plan your mobile marketing strategy based on how people use their mobiles; do some market research or perform a survey of clients' mobile habits: for example, a very common use is to locate a business that is local to where the person is located at the time, utilizing the GPS functionality of the mobile.

■ Strategy, interface design, and visual mobile site should allow for the discoverability and share-ability of the Web, while still allowing for the connected platform nature of an app.

The Mobile Media Matrix

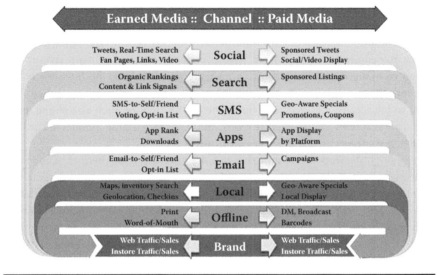

Figure 1.1   Mobile channels leading to a brand.

■ Consumers are getting into the habit of hunting down businesses on the move, looking for contacts, directions, product comparisons, price checks, and full-on purchases. So keep it simple. Put the most important information front and center, and make sure that everything on a mobile site absolutely needs to be there.

■ Mobile site developers are using JavaScript® *-based tools such as jQuery (<http//jquery.com>) to mimic the same-page interactivity that Adobe Flash made popular on desktop sites; it can help bring a mobile application to life without sacrificing simplicity.

■ Reducing the number of form fields is a good idea on any website, but in a mobile site, it is absolutely critical, and be careful with the credit card form: see Stripe & Square (<https://squareup.com>).

■ Finally, ask this important question—not so much technical as practical: Can the mobile site provide a benefit from having information at the moment of a purchase decision? Location, price, time of offer, etc.

---

* JavaScript if a registered trademark of Oracle Corporation.

1.8 Constructing a Mobile Site

The simplest way to construct a mobile site is to add JavaScript to an existing website that detects the kind of mobile device visitors are using and redirects them to the mobile site, on a sub-domain or separate mobile domain. The advantage here is that the enterprise or brand can have completely separate content, layout, and design to make it easy for visitors to navigate and consume the mobile-relevant content. The strategy is to rethink the site to appeal to mobile visitors and convert them to sales leads; it is critical to have contact information such as a phone number, a call to action, navigation directions, and maps for guiding mobiles toward the retailer, restaurant, brand, store, or company.

Although desktop and mobile sites use the same Internet, there are some basic differences in technology, aesthetics, and purpose that make them each quite different in terms of design and functionality. The mobiles allow users to do different things as they move, search, and share. A traditional website might include multiple videos, large images, and Flash animation, but that will not work with mobiles, due to their bandwidth and browser limitations—Apple does not support Flash at all. However, mobiles have their own unique functionalities; they can pinpoint a store's location within a few feet, and the user can tap on a phone number and be instantly connected.

Because traditional sites have been around for so long, a standard in design and layout has developed over time. Most people browsing such sites are already familiar with it—the eye naturally looks for navigation options at the top of the page, expecting the perimeter of the sites to be occupied by ads. However, mobile is a whole new frontier, so it is important to guide the user's eye naturally through designs and layouts more specific to mobiles. For example, having a full menu at the top of the page would not work for mobile sites, simply because of space. Mobile sites also cannot handle an overload of text like traditional websites can. Having less text that is larger lets users read comfortably and makes it much more likely that they will read it at all.

Due to increasingly larger monitors and faster bandwidth, traditional websites have grown to include dozens of sections with a lot of content. Mobile devices, however, have a limited viewing area, which means that it is important for each mobile page to have a single focus. Mobile sites should be linear and simple. Users want to interact with the page and be guided one step at a time. For example, if a retailer sells huaraches, their mobile site might have a list of the different types of sandals; once a user taps on a style, he or she can then narrow it down further by shoe size, brand, color, etc. Giving visitors a single task to complete on each page makes it easy for mobiles to navigate and a mobile site to succeed.

An analysis of mobile Web metrics can pinpoint where users are dropping off. With the right use of technology, design, and layout, a business will be able to draw in consumers to a mobile site that lets them tap and swipe through a user-friendly and interactive experience. The critical design feature is to *keep mobile users in mind at all times.* Remember that they are *moving,* meaning that the mobile site must be fast, quick, and to the point. A mobile site needs to load the homepage quickly, offering only a couple of clear choices specific to why the user is there. The layout must be designed for mobile users on-the-go; remember that mobile visitors are navigating with thumbs—with very small screens. A mobile site should be developed using the latest standards for HTML5 to minimize what the browser actually needs to do to load the page in order to keep it simple and fast.

The design should be clean and minimal; photos should be optimized, should be used only when necessary, and should be clean. The developer will want to utilize what is called "responsive design," which is the approach that suggests that mobile site design and development should respond to the user's behavior and environment based on screen size, platform, and orientation. The practice consists of a mix of flexible grids and layouts, images, and an intelligent use of CSS media queries of HTML that currently support media-dependent style sheets tailored for different *media types in different devices.*

Media queries extend the functionality of media types by allowing more precise labeling of style sheets. A media query consists of a media type and zero or more expressions that check for the conditions of particular *media features.* Among the media features that can be used in media queries are "width," "height," and "color" so that presentations can be tailored to a specific range of mobile devices without changing the content itself. For a definitive guide on CSS media queries, go to the W3C (<http://W3.org/TR/css3-mediaqueries>).

Navigation, which is the linking structure of a website, is dealt with much differently on a mobile site than on a traditional website. The buttons should be large and easy for the user to click, and the choices should be minimal and relevant. Again, remember that mobile visitors are moving. When visitors get to a mobile landing page, they need to see the main choices—with no more than two or three options—so they can select them right away. These choices need to relate to the reason that they are at that mobile site in the first place. If the user scanned a QR Code (Quick Response Code) or used an SMS (Short Message Service) shortcode (<http://usshortcodes.com/>), which is a number to which an SMS or text message can be sent (such as a mobile page), their decision to do this needs to be validated immediately; visitors should get to this page and know right away what to do. For example, an enterprise or a brand might consider using SMS to improve their marketing efforts by advertising to send a "text for information about a product or service to 12345" so that when users send this text, they receives one back with a link and maybe a short message directing them to video clips for that product or service at www.link.com. If a link is used when they click on it, they should end up on that mobile site's landing page.

Designing for a mobile site is much different than designing for a traditional website. You are dealing with thumbs, not fingers, so the minimum area required for an optimal mobile link is 44 pixels by 44 pixels. That is about the average area that a thumb requires to effectively select a link. If the mobile site links are very small and crowded together, visitors will have a hard time hitting the right link and will become very frustrated and quickly leave the site with a bad user experience, which make marketing and advertising impossible. Provide ample room for links to breathe and be effective. Ensure that visitors can easily select the links they want. This will require smart design and layout, using color and spatial arrangements to offer the user the easiest navigational structure possible.

Next, the text should be easy to read. Remember that mobiles are very small; developers must keep in mind that reading on such devices is more difficult than on a standard traditional website, and content should be kept to a minimum with fonts that are large and easy to read.

Responsive design really factors in here; the mobile site code can go a long way toward ensuring that the cross-platform user experience is optimal. The mobile site developer needs to ensure that the same visitor experience works, at a minimum on Apple and Android mobiles. As with any mobile marketing efforts, research, planning, defined goals, and extensive testing are required, keeping in mind the different versions of the iOS and Android operating systems. The mobile site developer should take the time and build a mobile experience that supports the enterprise and brand marketing goals. The ultimate goal of a mobile site is to capture mobile data for modeling and analytics.

Mobile site designers and developers must consider usability concepts about their mobile visitors. These considerations include *screen size, inline images, hyperlinks, font sizes,* and *page navigation.* Designing for mobile devices should be simpler than for a standard website and more *task based* to get the job done because these moving visitors are likely searching for something specific and urgent, such as looking for a nearby tow truck, a plumber, a garage, or a store.

Planning a strong user experience also means that the developer should consider how mobile visitors will be interacting with the site. On a traditional website, visitors can interact with a mouse and a keyboard, but on mobiles, they will be *tapping, flicking,* and *swiping* their way around the mobile site. So the design of the mobile site must consider how visitors can *easily and effectively access the site information with these physical motions and movements.*

The core of any site, whether wired or wireless, is of course the *page content.* In a carefully designed, planned, and tested site, each page should hold significant amounts of useful information for the visitors, such as text, photos, content, or videos. For the mobile site, however, this should be kept to a minimum because too many pages requires more loading and waiting time, which can kill the mobile visitor experience. Unless it is absolutely required, the mobile page content should be kept short. Short moving images if at all, and a single-column layout work well for mobile sites; the key for developers is simplicity. Mobile site developers should avoid tables and large images if possible because these tend to create download problems between different mobile devices.

Google AdWords™ now lets users search keywords that are specific to mobiles, helping to narrow down exactly what words the developer needs to use to reach the consumers. Use the same SEO techniques used for traditional website optimization with that of a mobile site—use these mobile keywords to create meta titles, title tags, and headers. As previously mentioned, do not overload mobile visitors; keep the content short and simple, and make sure images are smaller and important direction buttons are bigger.

Remember that this is all being seen on a small screen; realize that a mobile site is being viewed by users who may be strolling down the street reading their devices. Most sites have two URLs—one for mobile and one for the Internet; the SEO advantage to having a mobile site is that Google now has a designated bot that crawls around looking for mobile versions of traditional Internet sites to index. This means that both pages of the same site are indexed as one, a huge advantage because it means that if both sites are optimized properly, they are both equally searchable. If the mobile site is for a retailer, get an app developed that provides a competitive edge to keep customers from browsing away on the Web and looking at competitors; instead, it keeps visitors directly in the virtual store via the app.

Google provides a configuration tool to transcode a traditional site from classic HTML to mobile HTML but it might not give visitors a unified experience. Some of the problems are that the Google transcode conversion can result in having images resized in unsightly ways, duplicate content, error pages, and an overall bad visitor experience. To avoid this, the developer should consider making a sub-domain specifically for the mobile site. This is a key factor for search direction and indexing.

Having one distinct mobile URL keeps the mobile optimization from interfering with the traditional website optimization; it also keeps the same browsing experience on the small screen and allows the Googlebot-Mobile (<http://www.google.com/mobile>) to visit and index the mobile version for mobile searches. To accomplish this, avoid using Flash, Java, Ajax, and Frames; instead, try XHTML (WAP 2.0), cHTML (iMode), or WML (WAP 1.2). Developers should test their mobile sites by running it through Web Service Connector (WSC Mobile) (<http://validator.w3.org/mobile>; Figure 1.2) to ensure it is mobile friendly and test it on multiple browsers and devices.

Also, do not overwhelm mobile visitors with scrolling; instead, include "Previous" and "Next" buttons to help guide them through the content and webpages. Mobiles

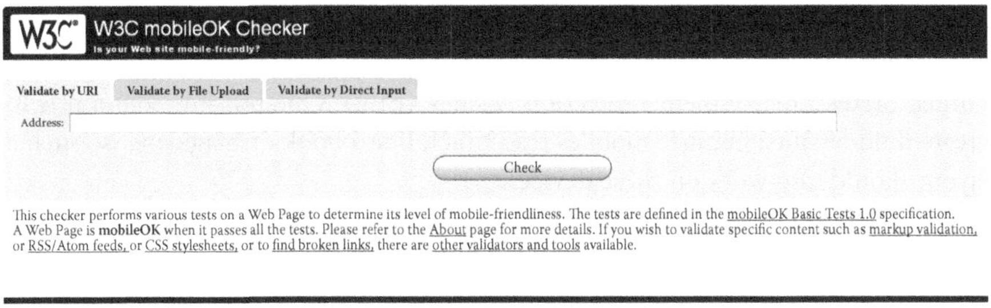

Figure 1.2   The W3C mobileOK Checker.

Table 1.2   Screen Resolution Size of Major Mobile Devices

Resolution Size	Mobiles
320×240	Blackberry®, Android, Symbian®
320×480	Android, Apple
480×360	Blackberry
360×640	Symbian
480×800	Android, Maemo®, Windows
768×1024	iPad®
640×960	iPhone 4®
1280×800	Android, Windows, Apple

might be getting smaller, but search opportunities and mobile consumer options are expanding—they now have indexing capabilities for products and services pricing—instant purchasing and sharing, and individually developed apps that remove them from the browser and place them directly inside virtual stores, thereby keeping their attention focused on the most important goal: conversion.

Remember that functionality is more important than style for mobile sites than for traditional websites; the biggest challenge is ensuring that the site looks the same and is compatible with multiple devices, so testing and validating this is vitally important. The mobile environment contains a rich variation of design considerations from different screen sizes and resolution to a variety of shapes. The developer needs to ascertain the specs of current mobiles because the goal is to display appropriately across a range of screen sizes without having to recreate pages for different platforms. Table 1.2 provides some screen resolution sizes of the major mobiles.

All of these resolutions are subject to change, depending on the model of the mobile, so extensive testing via WSC Mobile is required to ensure that the site displays properly under the operating system they support. This wide variety in display size makes it difficult to decide how to choose an appropriate layout size for mobiles, which have an average life of only around 18 months according to the United States Environment Protection Agency (EPA). One possible solution is to create fluid layouts; because mobiles read much like a book or magazine, so such a layout should also work on these devices.

Lengthy sections of text can be difficult to read, so placing them on several pages limits the scrolling and quickly gets mobile visitors to where they want to be. The mobile site developer should get rid of low-priority content. Stick to a single column of text that wraps so there is no horizontal scrolling. Good examples are the

Figure 1.3 CBS mobile site.

CBS (<http://www.cbs.com//mobile>; see Figure 1.3) and NBC (<http://www.nbc.com/mobile>) mobile sites that break up videos, shows, schedule, sports, and news articles into small portions from their main landing page.

Simplicity equates to usability, and mobile visitors should easily move around the site with no difficulty; therefore, avoid the inclusion of tables, frames, and other formatting because the more visitors have to click to links on a mobile site, the more they have to wait because of the loading time. The developer needs to strip down and simplify the mobile site with a balance between content and navigation. A good example is the mobile site of Best Buy (<http://m.bestbuy.com/m/b>), which lists only the most essential product categories, thus trimming down the level of hierarchy for content.

Through the use of mobile analytics and behavioral models, get to know what visitors are doing and be aware of what they are looking for. Find out how they will want to navigate the mobile site. The main purpose of constructing a mobile site is to aggregate important visitor behavior data for mobile analytics. It is important to clearly advise site visitors what items are available to them instantly. This can be done by changing the font and background color of links and buttons, or by simply adding some padding around links to make the clickable area larger at about 44 pixels by 44 pixels. Geek Squad® (<http://m.geeksquad.com/>) provides a good example of this strategy.

Each page download consumes time and system resources, the latter of which are in short supply for mobiles, so try not to force site visitors to dig or search through a multitude of pages in order to access the information they are seeking.

Figure 1.4   Note the simple links and mobiles that FedEx supports.

Aim for a balance between the number of links on each page and the depth of the mobile site. It is difficult to input text in mobile sites, so replace it with radio buttons instead so that visitors can quickly go where they want to go in a seamless, quick manner.

Developers should strive to make their content compelling enough and instantly usable by giving visitors what they want, when they want it; they do not want to be forced to dig deeper into the site. Aim for short URLs and simple choices, such as the FedEx® Mobile site (<http://m.fedex.com/mt/www.fedex. com>; see Figure 1.4).

Mobile sites can generate important sales leads for retailers and marketers in multiple ways, and here are just a few. Start by developing an email newsletter via text messages. Every mobile can now send text messages; take advantage of that to increase subscriptions via a simple text-message to an auto-responder and a mobile-friendly signup page. Ask visitors to text a particular word to a five-digit number; for example, invite visitors to text the word "INFO" to that five-digit number, called a *shortcode* because it is shorter than the normal full ten-digit phone number.

Many affordable SMSs can handle the text messaging side for the developer (e.g., TextMarks (<http://www.textmarks.com/>). When visitors text that "INFO" keyword, they will immediately receive a predefined reply message that contains a call-to-action or a sign-up page that is easy to read and use on any mobile; it should contain no more than one or two fields and communicate the benefit of joining the site. Use that mobile text message to create a mailing list.

Use the SMS to collect visitors' phone numbers, and their permission to get texts from the site; the advantage of SMS over email is immediacy. On average, an email is opened about six hours after it is sent, if at all. In contrast, some studies indicate that, on average, a text message is read within four minutes of receipt. In addition, email campaigns average a response rate of 17 percent, meaning the vast majority of them are never read; in contrast, SMS response rates average in excess of 95 percent. SMS has another advantage: the sign-up rates are much higher; they allow people to confirm immediately at the moment of receipt that they are most ready to join a mailing list and engage with a mobile site. The limitations of SMS marketing is that messages cannot exceed 160 characters of plaintext; however, it is a much more personal channel and should not be overused.

Also use QR codes to attract mobile site visitors. Almost all modern mobiles support the scanning of QR codes, which can be used to attract visitors via their browsers. The potential here is that someone can load a webpage quickly and easily even if it contains a long, complex tracking code. Most importantly, these codes can make it easy to attract and convert users via a simple scan of their device. Most mobiles require an installed, third-party app before they can read a barcode at all; however, that is changing as QR code scanning is growing extremely fast and will probably be a standard on all new mobiles by the end of 2012.

Finally, optimize mobile site landing pages, ensure that, at a minimum, the site works with Apple and Android mobiles the developer can use the mobile HTML5 boilerplate as a starting point. The tricky part here is testing on different devices and the hundreds of different models. One option for the developer not conversant with coding on HTML and CSS is to outsource it to a mobile site builder service such as Mofuse® (<http://mofuse.com>), Wapple® (<http:wapple.net>), or Atmio® (<http://atmio.com>).

Mobile site developers should test and measure the results of each of these techniques; they should ensure that third-party developers have experience in developing sites that meet these criteria. In the end, the mobile site developer may want to include both a QR Code and an SMS call-to-action to achieve a better response rate.

1.9 When to Build a Mobile Site

Currently, over 90 million people in the United States own mobiles, according to comScore (<http://www.comscore.com>); and if the trend continues—as most analysts and mobile vendors believe it will—the number of individuals in the United

States with mobiles will be close to, if not exceed, 100 million by this year. That is nearly one in three Americans. Who are these people, and what are they doing with these mobiles? They are consumers, customers, employees, and partners, and more than 40 percent of them are using their mobiles to browse the Web and shop online and download apps.

However, a majority of businesses and brands have failed to "mobilize" themselves, that is, to create a mobile site, or app. Does that mean that every business or organization needs a mobile website? No. But if a brand or enterprise currently has a business-to-consumer (B2C) or B2B digital presence, then it is time to develop a mobile strategy. The marketer or brand needs to ask the following questions:

■ Does the organization currently have a website that is regularly used by customers?
■ Do the people the brand is trying to reach use mobiles on a regular basis?
■ Can mobiles provide opportunities that a traditional Web presence, or other channels, cannot or does not do as well?
■ Would customers, employees, or partners benefit from having information at the moment of their moving decision?

If a "yes" to two or more of these questions came up, then the brand or enterprise *definitely* needs a mobile site.

Think of a mobile site as a real-time system of engagement—that is, a new way to improve the method of appealing to consumers, customers, employees, and partners. For example, for a real estate company, prior to mobile, if customers wanted information about a house, they would call the real estate agency or look up the information on their desktops. With mobile, however, the real estate agency can now provide prospective buyers with immediate information they need on their mobiles when they are right in front of the house. Mobile provides a new dimension of engagement and marketing that is more immediate, time and distance sensitive, and content relevant.

However, when selecting a mobile solution provider, the developer or marketer should go through the same vetting and request for proposal (RFP) process as they would for any other type of software or service, with part of the vetting process including viewing and testing several mobile sites versions, or apps, on a variety of mobile models. The key questions would include the following:

■ How is the user experience?
■ Does it have a good user interface?
■ Are the mobile site pages quick to load?
■ Is the site easy to navigate?

And finally and most importantly,

■ What are the analytics provided by the mobile solution provider?

Equally if not more important, developers or marketers should determine if the mobile solution provider can help them develop a mobile strategy, as opposed to just a mobile splash page or a basic app. Does the mobile solution provider have both the front-end design and user experience, along with the back-end integration expertise to make a mobile site truly a success for their business, enterprise, or brand?

One of the biggest mistakes organizations, marketers, and brands make when developing a mobile site or app is making it a stand-alone project. A mobile site or app should be integrated into their broader marketing, sales, and customer relation management (CRM) business intelligence overall strategy. Instead of just thinking mobile, they should think in terms of multi-channel, where mobile is just one channel or a new potent marketing component.

Developers and marketers need to understand their customers' goals and what mobiles they are using. So when creating a mobile website or app, make sure it looks good and is easy to navigate across a variety of devices and operating systems. Unlike traditional websites, with mobile it is all about streamlining information. So figure out what the five or six items are that will be the most vital to customers, and get rid of all the extraneous stuff that could slow them down or distract them, such as Flash, large graphics, or audio. Finally, make sure to test extensively the mobile site or app before releasing it.

Depending on the amount of work that must be done, it will likely take three to nine months to develop a good mobile site or native app—three months if the enterprise or brand already has a good service-oriented architecture in place and the mobile site or app is not too complex. As for the cost, expect to pay at least several hundred dollars to design and deploy a professional-looking, customized native app, although developers and marketers can find designers who will create a basic mobile site, with a few pages, for only a few hundred dollars.

But if the developer wants to create a multi-platform mobile presence that not only looks good on the front end, but provides a positive user experience and integrates with and leverages their enterprise's back-end systems, expect to pay several hundreds of dollars. While these expenditure may seem like a lot of money, when a brand or business considers that there are more than 100 million mobiles in the United States alone, and that that number is growing, the ROI (return on investment) can make a mobile site well worth it.

1.10 The Mobile Site Experience

Web browsers using their PCs expect, on average, a webpage to load within two to three seconds, after which they are more likely to abandon the website and move on. A KISSmetrics (<http://www.kissmetrics.com>) customer analytics survey in 2012 found mobile users still have a little more patience, and they need it. Almost three-quarters (73 percent) of participants cited slow mobile website page loads times as a problem they had encountered during the past 12 months. Findings revealed that over two-thirds (67 percent) of mobile Internet users expect page loads times on their mobile to take longer than on their PC.

Eleven percent said they expected load times to be much slower, 31 percent said a bit slower, and 25 percent said almost as fast as their desktop. The remaining 23 percent expected page load times on their mobile to be about equal (21 percent) to desktop speeds or faster than (11 percent) their PC experience. KISSmetrics asked mobile Web users just how long they were prepared to wait for a mobile site to load; most would wait between six and ten seconds (30 percent), while a fidgety (3 percent) would wait for less than one second.

There is little doubt that, as mobile Web browsing becomes more prevalent, expectations of fast page load times will increase. Mobile optimization is a must, and there are many tools, some free of charge, that enable mobile site speeds to be tested. By providing a satisfactory mobile experience, retailers pave the way for shoppers to go on to purchase via other channels. A ForeSee (<http://www.biz-report.com/2011/01/mobile-is-a-must-for-retailers.html>) research survey report found that 32 percent of mobile shoppers who were satisfied with their mobile experience were more likely to go on to purchase from that retailer online and 31 percent were more likely to buy product offline.

However, the ForeSee results warn that consumers are not concerned with the challenges faced when designing a mobile experience; instead, they do not want slow-loading and poorly formatted pages. In fact, by providing a satisfactory mobile experience, retailers pave the way for shoppers to go on to purchase via other channels; it is an easy and seamless way to acquire new consumers—if done right and having executed some of the guidelines in this book. The point is that a mobile site is only part of the total strategy for mobile analytics. The ForeSee survey found that mobile is a nice way to quantify customer touch points as they relate to a developer's or marketer's overall loyalty and sales strategy.

The ForeSee report found that retailers cannot afford to ignore or even neglect the mobile experience and assume it will not hurt their traditional online or in-store businesses. Retailers and brands need to make a good impression on consumers and keep them coming back to a mobile site by following these simple tips:

1. *Keep it simple STUPID:* For a mobile site, the less content is more. Cut out unnecessary images, and give visitors the option to look for what they want in a click.
2. *Cut down on clicks:* Make info a single or two clicks away to what they want.
3. *Make it finger friendly:* Users will be using their fingers to navigate, so make sure content and links are large and clear to make navigating the mobile site hassle-free.
4. *Test:* What looks great on one mobile may be a complete turn-off on another, so test the site on as many platforms as possible.

Performing a Web search from any mobile, users are likely to see completely different results than if they were to do the same search on a desktop or stationary device. This is because mobile and desktop searches are treated differently, with obvious differences such as location-based and time-sensitive information being prevalent in a mobile search.

1.11 Getting Mobile Googled

Traditional optimization techniques are still important, but understanding additional factors such as page and image size, text length, document type, model type, and operating systems can make a significant impact on the position a site receives in mobile search results. Because mobile users browse less, securing those top positions is even more important to mobile site developers. When it comes to setting up mobile sites, retailers and marketers need to make sure they create a specific sitemap that only includes mobile content URLs; more information can be found at <http://www.google.com/support/webmasters/bin/to pic.py?topic=8493>.

Essentially, Google states that a mobile sitemap can contain only URLs that serve mobile Web content, while any URLs that serve only non-mobile Web content will be ignored by the Google crawling mechanisms. This means that a developer who has any non-mobile content will need to create a separate sitemap for those URLs, and infers that if the <mobile:mobile/> tag is missing, the mobile URLs will not be properly crawled; however, URLs serving multiple markup languages can be listed in a single sitemap.

As Google is still the most popular search engine, retailers and brands need to understand exactly how it works and build their strategy accordingly. Essentially, Google uses two differing types of programs to search and index the Web: one for desktop searching via Googlebot and one for mobile sites via Googlebot-Mobile. Google currently does not crawl the Web with a specific algorithm for mobiles; however, it assumes that these devices are capable of producing a browser experience similar to desktops. This does not mean that brands should ignore mobiles in their marketing strategy, as any site that is going to be viewed by mobiles still needs to be optimized for mobile usage, and this includes using location-based keywords for the growing number of people performing localized searches.

Click-through rate and bounce rate tend to be a major factor in ranking mobile sites, especially for local searches. This may mean that brands and retailers need to look at different search keywords for mobiles to reflect the differing ways consumers search on these moving devices, and they may also want to look at a separate mobile strategy for paid search functions such as Google AdWords™.

Developers should take advantage of the Google Mobile Keyword Tool (<http://www.googlekeywordtool.com>) to find additional relevant keywords to make sure they have mobile optimized landing pages. It makes sense for content to also be formatted for mobiles as page size, load speed, and file types can all have an impact on how a mobile site is indexed. Another key area for mobile optimization is looking at how people get to the mobile content. To do all this properly, retailers and brands need to have clear insights into how their customers behave on the mobile site and how they are using it.

Mobile users, especially for retailers, can be at a completely different stage of the buying process as they may have already researched on a desktop and are just looking to complete the process quickly on their mobiles. They could even be in a store having found the item they want and be doing some price comparison. This means that the keywords and search experience are very different. The most common mistakes in mobile search arise when brands and retailers do not understand their customers' behaviors, needs, locations, interests, and goals.

Understanding customer behaviors in any channel is really about data capture and inductive and deductive analysis via modeling. The process of understanding mobile usage is the same as understanding website usage: ask them, watch them, test them, and analyze the behavioral data. At the very minimum, mobile developers should start to understand visitor basic usage information. Some free and easy ways to start building a picture of how visitors are behaving on a mobile site include looking at analytics reports and using Google Analytics™ for Mobile (<http://code.google.com/apis/analytics/docs/mobile/mobileWebsites.html>) or Google Analytics Mobile Apps SDK (<http://code.google.com/apis/analytics/docs/mobile/download.html>).

With most mobile search being about localization, retailers and merchants need to make sure that they keep any localized search sites fully up-to-date. Developers need to make sure that things such as Google Places™ and Bing® Business Portal are up-to-date; they need to add contact details and promotions, as well as encourage reviews and visitor interactions. Retailers and merchants should also look at signing up for local directories such as LocalDataSearch (<http://www.localdatasearch.com>), Yelp (<http://www.yelp.com>), Frommer's (<http://www.frommers.com>), Qype (<http://www.qype.co.uk>), and Tipped (<http://www.tipped.co.uk>). The following provides ten vital tips that will allow retailers, brands, developers, and marketers to create informed mobile search strategies and allow them to make their mobile presence as effective as possible:

1. Follow the guidelines of best practice for Web design and traditional SEO.
2. Create specific keywords for mobile searching based around locations.
3. Understand how the different mobile devices work with the mobile site.
4. Understand at what stage visitors are on the purchase journey.
5. When creating content for the site, keep it short and concise.
6. Understand how search engines index mobile sites.
7. Make sure all local details and directions are correct.
8. Understand how visitors are using the mobile site.

9. Look at how visitors get the mobile content.
10. Create a mobile site map.

1.12 Mobile Is about Now!

Web content is being consumed via mobiles more than ever before—consumers no longer want to wait for information. They want it in the car, in line for morning coffee, on the sidelines at soccer practice, while shopping, and between classes. The popularization of mobiles has transferred the online experience from the desktop to the pocket, purse, and backpack. The result is a world in which mobile is quickly becoming the most important medium for information transfer. It is essential that all mobile site developers reliant upon their Internet presence adapt quickly and effectively.

Mobile site developers need to make sure their content management system allows them to deliver content based upon screen resolution—not just specific devices. Being able to detect screen resolution is often more important than detecting a particular mobile. In order to provide users with the best experience possible, developers need to know how much screen real estate they have available. They need to identify the most common screen resolutions and create content that will fit well on most mobiles.

In most cases, the navigation structure on a website is not going to work well on a mobile site. The navigation experience on a mobile site needs to be optimized for touch input. The design and layout of the site menu should also be optimized for viewing on a small screen. There is nothing wrong with the front page of the mobile site just being navigation links. Test the navigation of the site on Apple and Android mobiles—not an emulator. An emulator still requires the developer to interact with their site using a mouse, which is much different than holding a mobile in a human hand and trying to click on a link with a finger.

Create key simple direct categories; for example, for a real estate mobile site, this might be (1) Easy property search, (2) View nearby properties, (3) Mortgage calculator, (4) GPS search, etc. Most mobiles make it easy to access the current geographic location of the device and its owner. The developer needs to leverage this capability to make it easy for users to get directions to the nearest office location. For example, for the real estate company, this would allow users to locate nearby properties based on their current location with a "GPS search" mobile menu option.

While some mobiles will automatically detect phone numbers and email addresses, the mobile site developer needs to make sure to create clickable links whenever possible. Mobiles make it easy to click to call, click to text, and locate addresses on a map. The developer needs to leverage these uniquely mobile capabilities to create a better experience for the site visitors who are moving toward the purchase of some product or service and make the visit immediately gratifying.

Google recently completed a survey, with the help of the research firm Ipsos (<http://www.ipsos.com/>), about how consumers use their mobiles; the findings underscore why businesses and marketers should care about this market channel. Mobile ownership has reached 44 percent of the total population in Spain and 38 percent in the United States. Mobile owners use their devices a lot, with 93 percent of U.S. users reporting daily usage. Better still, the majority of mobile users—88 percent of users in the United States—say they notice and respond to online ads. Mobile device users are also avid users of online video, mobile apps, and social networking services. Finally, over one-third of U.S. mobile users have purchased products or services over the Internet with their devices—an activity that is expected to become more prevalent.

If the survey statistics are not compelling enough, Google has released some new tools to help marketers and businesses create mobile sites. These include the GoMoMeter (<http://www.howtogomo.com/en/d/test-your-site/#gomometer>), which shows how a site appears in a mobile browser; the site also includes a list of companies that develop mobile sites, that is, guides to mobile ad agencies, advertisers, and publishers. Google predicts that one million small businesses around the world will build a mobile site this year.

1.13 Mobile Payments

With increasing mobile adoption and with users in turn becoming more and more comfortable using their handheld devices to do things traditionally reserved for the Internet or desktops, today every industry and marketplace is going mobile. But when you consider the effect that mobile technology has yet to make on transactions, there is a huge opportunity for disruption. Carriers, OEMs (Original Equipment Manufacturers), start-ups—everyone knows this; however, within a couple of years, mobile payment solutions have the potential to be a massive business. Yet, as things stand right now, carriers and mobile OSes are still finding it difficult to come to terms, with users suffering as a result. That is why SCVNGR (<http://www.scvngr.com/>) decided to spin-off LevelUp (<http://www.thelevelup.com/>), a mobile payment and rewards network.

The mobile payments network has been growing fast; it now counts over 100,000 users and has partnered with 1,400 merchants in Boston, New York City, Philadelphia, Atlanta, Seattle, San Francisco, San Diego, and Chicago. First and foremost, the problem that LevelUp is really trying to solve is offering a way for users to pay for their goods via their mobiles. Users simply use the LevelUp app to register their most-used credit card or debit card on the payment network; from there, they get their own unique QR code, which they can simply scan at one of those 1,400 merchant locations. The transaction is completed, and they are emailed a receipt.

Simple enough, right? But the real key to success in the mobile payments game is scale. It is about offering a solution that is carrier- and credit card-agnostic. ISIS (<http://techcrunch.com/2012/02/27/isis-revealed-carrier-led-mobile-payments-venture-shows-off-its-new-app-announces-banking-partners/>) has the clout of all the major carriers and credit card companies; ISIS is the carrier-led joint venture between AT&T, T-Mobile, and Verizon for payment processing for Chase™, Capital One™, and Barclaycard™. Under the terms of the deal, the banks will include their debit, credit, and prepaid cards into ISIS' forthcoming mobile wallet.

However, LevelUp is attempting to do away with the limitations imposed on mobile payments solutions by those typical niche requirements, so that anyone with a Web-connected mobile can show up at his or her favorite merchants and pay. LevelUp has released an API (application programming interface) that will allow developers to accept LevelUp payments from third-party software or POS (point-of-sale) platforms.

On the other hand, Sprint® has teamed up with Google for their Google Wallet™, in which both have their own agenda for the future of mobile payments. When you look broadly at mobile payment technologies, you have PayPal™ (<https://www.paypal.com/>), Square (<https://squareup.com/>), Amex Serve™ (<https://www304.americanexpress.com/credit-card/>), Visa PayWave® (<http://usa.visa.com/personal/cards/paywave/index.html>), Stripe (<https://stripe.com>), Master Card PayPass® (<http://www.mastercard.us/paypass.html#/home/>), and Dwolla (<https://www.dwolla.com>)—all in various ways building payment solutions around certain aspects of mobile transactions via mobile sites and apps.

1.14 Adobe Shadow™ Mobile Site Tester

The Adobe Labs preview of Shadow makes it possible to browse mobile Web content on a PC, or on other mobiles running iOS or Android operating systems. *Shadow lets mobile site developers synchronize browsing between desktops and multiple mobiles to check content displays across different platforms.* Adobe Shadow solves one of the biggest problems in mobile site development: figuring out what is broken on a mobile site and why. If a mobile site running HTML, CSS, and JavaScript looks great on the PC and an Apple mobile but horrible on an Android mobile, developers can use the Google Chrome™ browser's built-in debugging tools to tweak their code until they have achieved cross-platform perfection.

Shadow is an inspection and preview tool that allows front-end Web developers and designers to work faster and more efficiently by streamlining the preview process, thus making it easier to customize websites for mobile devices. Shadow will be updated regularly by Adobe to stay ahead of Web standards, Web browser updates, and support for new mobile devices entering the market, while incorporating user feedback to provide the best possible functionality and experience.

To make this work, a developer must pair the mobiles with the Adobe Shadow application, somewhat similar to pairing a Bluetooth headset with a mobile. The developer needs to enter a code generated by the mobile into the application to confirm that Shadow is authorized to pilot the device by remote control. From that point until the connection is broken, the developer can open a webpage in Chrome and a mobile—or potentially a whole workbench full of mobiles—to follow along. The developer can also pick up any of the mobiles and interact with them directly, say, to check some screen interaction, and then resume synchronized browsing to another webpage or application.

Adobe Shadow is important for resolving the differences in the ways mobile sites and apps work and display differently on different mobile browsers. Mobile developers are really struggling today, because of the demands of their clients, to test and develop mobile sites quickly using standards-compliant code. Shadow allows developers to use the code inspector in Chrome and see changes displayed on the paired mobiles. Adobe chose Chrome because it has the best and most popular code inspection and debugging tools for Android mobiles and the most aggressive implementation schedule for the open-source WebKit (<http://www.webkit.org/>) browser engine for Apple mobiles; the product also makes use of Winre, an open-source remote code inspector.

Adobe does something similar for Web development with Adobe BrowserLab™ (<http://browserlab.adobe.com/>), which simulates how Web content will be displayed across a variety of mobiles, but Shadow lets developers see their content on the actual mobile browsers. The biggest thing for developers is that when they are troubleshooting on desktop browsers, these are tools they can use to pinpoint the piece in the code that is the issue, and Shadow now allows the same debugging ability for mobile browsers. The Shadow app requires at least an Apple iOS 4.2 or Google Android 2.2 operating system. In testing, Shadow has been able to control more than thirty devices simultaneously, which is handy for testing from multiple manufacturers.

1.15 Mobile Cookies

Mobile analytics is about tracking device behaviors of mobile site visitors in a similar way to traditional Web analytics. In a commercial context, mobile analytics refers to the use of data collected as visitors access a site from mobiles. It helps to determine which aspects of the site work best for mobile traffic and which mobile marketing campaigns work best for an enterprise or brand—including mobile advertising, mobile search marketing, text campaigns—and the promotion of mobile sites and services via the use of mobile cookies or beacons.

Today, most mobiles are cookie enabled; companies such as Umber Systems are working to bring cookie-type solutions to standard mobiles. Data collected as part of mobile analytics typically includes page views, number of visits, number of visitors, and countries of origin, as well as information specific to mobile devices, such as device model, manufacturer, screen resolution, device capabilities, service provider, and preferred user language.

This data is typically compared against key performance indicators (KPIs) for mobile site performance and ROI, and is used to improve a mobile site's performance and mobile marketing campaign's response rate. The majority of modern mobiles are able to browse websites, some with browsing experiences similar to those of stationary computers. The W3C Mobile Web Initiative (<http://www.w3c.org/Mobile>) identifies best practices to help websites support mobile device browsing.

Many enterprises, marketers, and brands use these guidelines and mobile-specific code such as Wireless Markup Language (WML) or HTML5 to optimize websites for viewing on mobile devices. Consumers are their devices—as with traditional websites, cookies and beacons are important Internet tracking mechanisms—that customize the visitor experience by remembering the visitors' preferences and behaviors. The same holds true with mobile sites, where these tracking mechanisms can perform the same personalization functions.

These mechanisms are a reliable and proven method of aggregating the type of behavioral data that is ideal for mobile analytics and marketing. Unfortunately, what is universally acceptable on the Internet with stationary devices does not function with some mobiles. For a best practice guide to creating and utilizing mobile cookies for delivering relevant content to mobiles, the developer should go to the mobile section of the W3C (<http://www/w3/org/TR/2008/REC-mobile-bp/-20080729>). This site specifies the best techniques and technologies for delivering Web content to mobile devices, including cookies and other tracking mechanisms.

As with the Internet, cookies are frequently used to carry out session management, to identify users, and to store user preferences. Many mobiles, however, do not implement cookies or offer only an incomplete implementation of this function. In addition, some gateways strip cookies while others simulate cookies on behalf of mobiles. Without a doubt, the use of cookies is far more complicated when targeting mobiles and the hundreds of models and different operating systems.

The W3C recommends that developers test to see if cookies are supported by the specific mobile on its current access path; it also suggests that if cookies are not supported by certain mobiles, that developers use a uniform resource identifier (URI) (<http://www.w3.org/Addressing/URL/URI_Overview.html>) decoration for session management, being careful not to exceed the device's maximum length for such strings. A URI is a string of characters used to identify an abstract or physical resource on the Internet.

Such URI identification enables interaction with representations of the resource over the Internet using specific protocols. Schemes specifying a concrete syntax and associated protocols define each URI. In addition, some gateways provide user identification without the setting of mobile cookies. The principal objective is to improve the user experience on the Web when accessed from mobiles. The W3C recommendations refer to delivered content and not to the processes by which it is created, nor to the devices or user agents to which it is delivered. It is primarily directed at creators, maintainers, and operators of mobile sites, and offers some cookies requirements, delivery context, structure formats, and conformance for mobile devices.

There have been privacy concerns with Internet tracking mechanisms such as cookies, but the concern that they are used to target individuals is actually misplaced—because they are really tracking *mobiles*. These mechanisms allow for mobile sites to approximate consumers' preferences and for letting marketers, brands, and enterprises improve their responses to customer demands for relevancy in the content they provide, as well as the products and services they offer.

Mobile sites are a relatively recent channel for accessing the Internet, marketing, and advertising; they introduce a new layer of complexity that can be difficult for websites to accommodate because they require *cross-platform* functionality and diminutive displays. For example, Apple mobiles do not even accept third-party cookies, which are served by such ad networks as DoubleClick™ (<http://google.com/doubleclick>). The ability to target and segment customers using mobile analytics demonstrates the strategic desire to improve customer loyalty and advocacy.

The mobile site developer may also want to outsource the use of first-party cookies to such firms as TruEffect (<http://trueffect.com/>); their platform enables the serving of mobile cookies and ads directly from an enterprise's domain rather than the ad server's domain such as DoubleClick or Microsoft Advertising Atlas (<http://www.atlassolutions.com/>). This enables their clients to use their own customer data to place relevant content that is determined by the customer viewing the ad placed by the TruEffect cookie. This enhances segmentation by building deeper insight into finding prospects who behave like customers, not simply matching them demographically.

1.16 The ConnectMe QR™ Network

ConnectMe QR (<http://www.connectmeqr.com/?a=492>) provides QR Codes and editable mobile website marketing solutions for developers. Their service provides QR code for mobile marketing, including their mCard service, which is an around-the-clock, editable service that developers can use to control their mobile site content. ConnectMe QR offers a turnkey solution designed especially for mobile site developers.

This mobile marketing company also provides their new Advanced Mobile Website Editor along with their QR code subscription service. When they introduced their flagship product, the mCard, their focus was on pointing to subscribers' social media and contact information. Now, with their Advanced Editor, users can customize their mCards and their mobile landing pages along with their ConnectMe QR code points by selecting from over 200 icons and entering their own text. These icons are links to information a subscriber wants someone to see when they scan their ConnectMe QR code. It can contain up to nine links such as phone numbers, email addresses, YouTube videos, special coupons and offers, social media accounts, interactive location-based maps, etc.—it is completely up to the subscriber.

Network marketing companies and mobile site developers, or any group of visitors who share a common goal or interest, can improve their social communication within their circle. ConnectMe QR has created the perfect tool for groups with similar collective interests. Their Advanced Editor Mobile Platform can link and keep groups engaged and connected with immediate and relevant information. It is an easy-to-use service that is cost effective and also an easy-to-use marketing tool for any developer or marketer.

ConnectMe QR offers mobile site developers the ability to post and distribute videos, and update sales tools and printed materials that communicate the latest pricing, information, and incentives (Figure 1.5). QR codes must do more than link to a single coupon or website; their mCard provides a mobile site to be a more effective and simple way to control their mobile marketing efforts. Within minutes, developers can update their sales tools, videos, agendas, event calendars, and/ or contact information. They can provide interactive, multimedia, and marketing information readily available to prospects and visitors, no matter where they are at the moment. The mCard service is offered via a thirty-day free trial; once the trial ends, the cost is currently $4.99 per month without any contract.

Figure 1.5   ConnectMe QR™ and customizable mobile site service.

1.17 Mobile Ad Networks

A mobile ad network is an Internet platform provided by a media company, on which marketers advertise their mobile site and their products and services on a publisher's webpage across to mobiles. One benefit of ads placed on a mobile network is that consumer distribution is highly targeted. This means that ads reach an audience that is most likely to convert or become customers of the product or service being marketed via mobiles based on specific user criteria, such as age and gender. These networks have different business models, geographical coverage, publishers, and advertisers, with different targeting capabilities, pricing, sales models, etc., that the mobile site developer will need to consider before choosing the right one for them.

At this point, the mobile advertising market is so saturated that mobile site developers and marketers have many options for where and from whom and how to buy ad space for their mobile site, apps, products, and services. There are global ad networks from the major mobile manufactures such as Google AdMob™ (see Figure 1.6; <http://www.admob.com/?) and Apple iAd® (<http://advertising.apple.com/?cid=wwa-naus-seg-iad100604-0000048%cp=brand&sr=sem>).

There are also some independent players, such as Millennial Media (<http://millenialmedia.com/>), Jumptap (<http://<ww.jumptap.com>; Figure 1.7), and InMobi (<http://www.inmobi.com/>). There are also mobile demand-side ad

Figure 1.6 Google AdMob™ mobile advertising service.

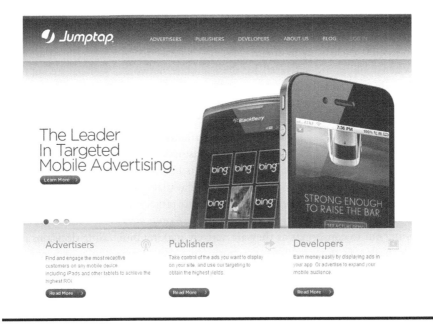

Figure 1.7   The Jumptap portal for mobile developers.

network platforms, such as MdotM and Fiksu and conventional demand-side platforms, such as DataXu (<http://www.dataxu.com>), Turn (<http://www.turn.com>), and Invite Media (<http://www.invitemedia.com>).

However, with so many different mobiles and an ever-expanding number of new models and upgraded operating systems, it is difficult to build advertisements that will be effective; but new standardizations are evolving, such as the Interactive Advertising Bureau (IAB) Mobile Rich Media Ad Interface Definitions (MRAID) (<http://iab.net/mraid>).

As the mobile space begins to mature, developers, marketers, and publishers are starting to look for new ways to monetize beyond the single ad network system. For example, the gaming giant ngmoco (<http://www.ngmoco.com/>) has adopted ad serving and operations and sophisticated reporting platforms to power advertising on their mobile site and apps. Companies such as Backflip Studios (<http://www.backflipstudios.com/>) already have established in-house ad operations teams specifically focused on mobile sites and apps, and this number will expand.

Then there are mobile networks, which are different from ad networks because they know *who* consumers are and *where* they are. This is something that is absolutely necessary so they do the job they were created to do: reach devices via voice calls and text message—any place in the world. Mobile networks also differ from the Internet in that they are privately owned, such as Sprint (<http://www.sprint.com/>), Verizon Communications (<http://adage.com/directory/

verizon-communications/289>), AT&T, and T-Mobile (<http://t-mobile.com/>); and because they are privately owned, the endpoints of the network are controlled by, and visible exclusively by them.

When mobiles are activated by these mobile networks, owners are supplied with a unique number that can be directly associated with an individual consumer. These mobiles can also be located geographically by triangulation between cell towers, or with even greater precision using GPS and Wi-Fi, or a combination of the two. So the future of mobile analytics and marketing may lead to a combination of ad *and* mobile networks triangulating on consumers by way of their mobiles. The main reason mobile advertising is exploding is because mobile ads are *highly targetable;* enormous amounts of information are available about mobile device users in determining exactly which demographics to target via location and interests. Most importantly, mobile site marketers *can reach consumers at the point of purchase.* Reaching out to consumers as they are actively making purchasing decisions gives mobile marketers an incredible opportunity to convert them, while they are roaming and browsing along their shopping journey.

Mobile developers, marketers, and brands have two key methods of reaching consumers and their devices: Is it more efficient to drive users to visit their mobile site or persuade them to install their mobile app? Then there is the issue of deciding which mobile ad network to sign on with. Mobile site developers need to consider what type of mobile traffic they are looking for: Are they just building their brand, or are they looking for more clicks to the mobile site based on GPS proximity, Wi-Fi triangulation, or keyword searches?

1.18 Mobile Site Developers

There are many developers specializing in the creation of mobile sites, which enterprises, marketers, and brands can outsource for creating a mobile site; the following are some of the best in this rapidly evolving industry.

■ Blue Corona (<http://www.bluecorona.com/>): Offers mobile sites, marketing analytics, and mobile metrics. Blue Corona can construct a site optimized *specifically* for mobiles with prominent company logo and contact information with click-to-call functionality and direction via Google Maps; they also provide metrics via Google Analytics Web analytics service.

■ Phonify (<http://www.phonify.com/>): Constructs mobile sites geared for shoppers on-the-go. Mobiles are always moving and want to compare prices and find the nearest store location; they embed Google Maps for that. They add a small block of JavaScript on a client's main website to automatically redirect mobile traffic; they offer three price plans with hosting options with click-to-call or click-to-email functionality (Figure 1.8).

Figure 1.8 Mobile site creations (Phonify Inc.) to reach the moving marketplace.

- Mozeo (<http://www.mozeo.com/>): Provides mobile website construction, text messaging, and advertising services (Figure 1.9). Clients have full control of the style, look and feel, and the content of their mobile site. This developer uses their mobile sites to complement marketing campaigns with rich media optimized for mobiles. They provide their clients with QR code capability and geo-location for mobile marketing campaigns. They can create a shopping catalog and carts with full credit card capabilities and mobile-rich media. They give clients the ability to take advantage of these features in a completely integrated environment.
- mobiSiteGalore (<http://www.mobisitegalore.com/>): Users can create free mobile sites in minutes; hosting is available. Their Wireless Application Protocol (WAP) sites are built with their proprietary builder, which are guaranteed to work on any mobile. This developer claims that no technical knowledge, programming skill, or Web design experience is required to build a site with their WAP builder. There is no software to download or install and no plug-ins are required either. A fully integrated HTML richtext editor helps clients create rich XHTML content with text and images in a WYSIWYG mode with absolutely no need for prior knowledge or experience in HTML or XHTML. Clients can choose from a range of ready-made design templates

Figure 1.9 **Mozeo offers text messaging and mobile site creation services.**

with customized colors, fonts, and layouts that are in 100 percent compliance with W3C Mobile Web Standards.

■ Mobile Web Up (<http://mobilewebup.com/>): Provide sites optimized for mobile devices; offer free evaluations. This developer can create mobile sites designs, hosting, marketing, and advertising.

■ Mobify (<htttp://mobify.com/>): Provide a mobile website studio as well as support for mobile ads and apps design and construction. Mobify uses HTML5, CSS, and JavaScript to fully integrate with traditional websites. Mobify mobile sites take advantage of touch-based navigation and widgets such as tactile image carousels, enabling users to tap and swipe their way through, say, a store—with a finger-friendly interface and elements that resize based on device specifications and capabilities.

■ Usablenet (<http://wwwusablenet.com/>): Platform requires no client IT resources, systems integration, or Web development. All mobiles and all types of output are supported. There is no aspect of their mobile sites that they cannot transform; their development platform is crafted and refined to support their clients' objectives and all types of mobile devices (Figure 1.10).

■ Digby (<http://www.digby.com/>): Platform enables mobile site and rich apps for all major mobiles. They make remote storefronts available to all types of mobiles anytime, anywhere. They make virtual sales representatives—with access to product information, reviews and promotions—which complement a retailer's catalog with up-to-date product information, videos, and reviews.

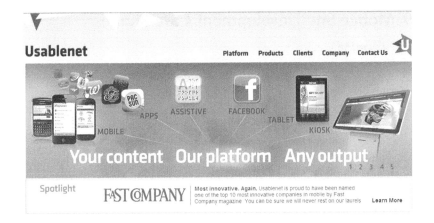

Figure 1.10   Usablenet was recognized by *Fast Company Magazine* (March 2011).

▪ DudaMobile (<htttp://www.dudamobile.com/>): Is listed in Google GoMo, with an average price for a site between $200 and $500 with hosting and analytics included. Clients can enter their traditional website URL and instantly get it converted, built, and tested into a mobile site, which is "thumb-friendly" and intuitive with lots of big, fat buttons and "mobile-friendly" font sizes (Figure 1.11).

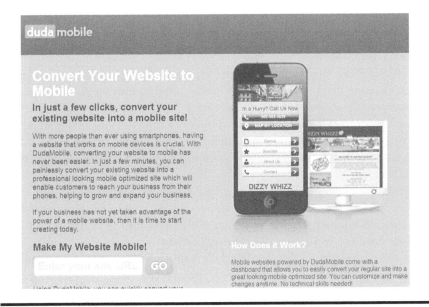

Figure 1.11   DudaMobile converts a traditional URL into a mobile site.

1.19 Mobile Site Development Case Studies

1.19.1 OneIMS

OneIMS (<http://www.oneims.com/>) is an online marketing company that offers Web design services specifically developed for mobiles. While standard websites can be viewed on mobile devices, their layouts do not always offer a comprehensive or user-friendly viewing experience. OneIMS hopes to change that by offering Web design and development tailored specifically to mobiles.

More and more people are doing the majority of their Web browsing, and even their online shopping, on mobiles. OneIMS is offering mobile Web design due to a reflection of a changing standard in the way people do business online. This developer offers full compatibility for all website features, including shopping, blogging, and more.

OneIMS clients with mobile sites will be better able to reach busy customers through mobile marketing and other techniques that are exclusive to mobiles. While mobile sites will differ slightly in terms of layout, they will provide the same services available on traditional websites. Their goal is to make the mobile sites as similar as possible to the original, but with a better user interface. The mobile sites are formatted for touch-screen devices, meaning that customers looking through their mobile sites will be able to enlarge and scroll smoothly without the glitches that can occur on websites designed for conventional viewing.

1.19.2 YoMobi

YoMobi (<http://www.yomobi.com/>) offers a free "Do-It-Yourself Mobile Website Builder" platform. The company designed their platform specifically for small businesses, enabling them to rapidly create mobile sites, compatible with all mobiles. Set-up, creation, and implementation of their mobile sites are free. YoMobi is providing developers to create their own advanced, tech-savvy mobile site. Most sole entrepreneurs do not have the time, money, or staff needed to customize mobile sites, and most mobile webpages are difficult to navigate, often turning away potential customers.

The YoMobi no-programming-required platform provides users with a fill-in-the-blank template, effectively rendering a mobile website builder comparable with desktop counterparts, but more accessible for mobiles. Through this service, YoMobi does not require small businesses to have an established website, thus allowing anyone to quickly go mobile, first. YoMobi is a free platform that allows developers to quickly build interactive, social-media-ready, mobile-optimized sites. The company's mission is to remove the barriers of time, money, and skill that often impede small business owners from quickly adopting new, revenue-producing technologies.

1.19.3 MobiTily.com

MobiTily Technologies (<htttp://www.mobitily.com/>), a mobile application development company, offers multiple mobile website themes for designers who can refer to these themes and customize them as per their business needs in the fastest possible time. MobiTily understands the trend that mobile sites are here to stay, as everybody now and in future will be going mobile in performing a lot of Web-related tasks. To capitalize on this euphoria, more and more business owners are looking forward to mobile versions of their websites, and are equally interested in mobile application development as they want to penetrate the enormous market potential of millions of mobile Internet users.

MobiTily mobile website templates are based on a combination of HTML and CSS, and the available themes can be used for various domains, including Entertainment, Tourism, Consumer Products, Automobiles, Real Estate, and other Services for the B2B and B2C sectors. These mobile website themes can be modified further to make a customized mobile website; MobiTily mobile website themes are available for free download.

Finally, most of these developers support the construction of mobile sites and apps; depending on the brand and business, one or both should be developed, but the mobile site is of paramount importance and should have priority. Most consumers are going to search for a business via their mobiles, with the option of downloading an app from them after they visit the site. A study by Morgan Stanley (Morgan Stanley.com 2012) suggested that by 2015, a large percentage of those surfing the Internet will be using mobiles.

1.20 Tracking via Unique Device ID

Apple made a decision to deprecate the Unique Device ID (UDID) for their iOS mobiles, and that decision has sent the mobile advertising and analytics industry into a search for new tracking techniques. Mobile analytics and tracking services relied on the UDID for information about users' behavior and engagement. The UDID allowed developers, trackers, and advertisers to send personalized push notifications, improve conversion rates, and engage users. Developers are now coming up with ways to get around the UDID issue.

Velti (<http://www.velti.com/>), a company that provides mobile marketing and advertising technology, has formed a working group dubbed ODIN (Open Device Identification Number) with leading analytics, marketing, and advertising services such as Jumptap (<http://www.jumptap.com/>), mdotm (<http://mdotm.com/>), Smaato (<http://www.smaato.com/>), RadiumOne (<http://www.radiumone.com/>), StrikeAd (<http://www.strikead.com/>), Adfonic (<http://adfonic.com/>), and SAY Media (<http://www.saymedia.com/>). The goal of the ODIN Working Group is to develop an alternative secure anonymous device identifier for the mobile advertising industry. The current ODIN solution creates an ID derived from the MAC address and is obfuscated to protect user privacy.

ODIN is intended to be a number designed to uniquely identify users' mobile devices in a convenient, interoperable manner, and the goal for ODIN is to be anonymous, consistent, and secure. The working group contends that anonymous device identifiers benefit the mobile ecosystem because they allow for anonymous personalization, ad frequency capping, and verification of application installations. The ODIN Working Group is actively working with leading privacy organizations, including the Digital Advertising Alliance (<http://www.aboutads.info/>), TrustE (<http://connect.truste.com/>), and Evidon (<http://www.evidon.com/>), to ensure the highest levels of consumer privacy.

Opera Software launched App-Tribute (<http://www.opera.com/press/releases/2012/04/03/>), a solution that allows mobile publishers and advertisers to receive marketing and analytics data without taking sensitive data elements such as unique UDIDs, cookies, or MAC addresses. Opera launched App-Tribute via its advertising subsidiaries AdMarvel (<http://www.admarvel.com/>), Mobile Theory (<http://mobiletheory.com/>), and 4th Screen Advertising (<http://www.4th-screen.com/>). App-Tribute is available for iOS and Android mobiles, and supports app-based advertising campaigns and enables cost-per-click (CPC), cost-per-download (CPM), cost-per-install (CPI) and cost-per-mille or 1,000 impressions (CPM) promotions.

For app developers, App-Tribute provides feedback to determine which publisher app should be attributed for promoting an app download and installation. App-Tribute consists of an App-Tribute Advertiser SDK that developers include in their applications to track successful downloads and subsequent installs of an app, as well as the App-Tribute Publisher SDK, which tracks the promotion of apps and anonymous consumer interest in designated apps. ODIN creates an ID from the Media Access Control (MAC) address and obfuscates it to protect users from being personally identified.

ODIN is designed to identify user mobiles in a way that is convenient to the advertising and analytics companies. It is supposed to be anonymous and consistent regardless of operating system, and is transported securely. Some argue that a hashed MAC address is no better than a UDID because it cannot be erased the way a cookie can on a browser. Advertising companies and similar monetization platforms used the UDID to track user download behavior by saving the UDID in their database and cross-referencing it with all the apps that support the monetization platform. In this way it is possible to find out the user's choice of apps, their location, and more, to improve their conversion rate.

ODIN is not the only alternative; there are also HTML5 cookies for tracking hybrid apps; SecureUDID (<http://www.secureudid.org/>) from Crashlytics (<http://crashlytics.com/>) is another. Then there is Core Foundation Universally Unique Identifier (CFUDID), an ID that is individual to every app, even when

Figure 1.12   Nexage, an advertising exchange for tracking mobiles.

apps are on the same mobile. The mobile real-time bidding exchange platform Nexage (<http://nexage.com/>) outlines options for the industry using that mechanism (Figure 1.12).

In advertising solutions that create a tractable ID, a user clicks through to the advertised app, which is later matched to an ID created upon its download and is used primarily for conversion tracking and attribution. In in-app solutions, where the app download generates a cookie by launching the browser upon download, the mobile cookie is the ID and is also used for conversion tracking.

Next, in device fingerprinting, a unique mobile identifier is created using attributes for each device. Using a hashed version of the device MAC address to replace UDID, or various other open-source solutions such as OpenUDID and ODIN, unlike the Web, mobile advertising does not work through cookies. That is why some type of ID recognition is extremely important to the mobile ad industry.

1.21 Mobile Site versus Apps

A report by Nielsen's analysis (Nielsen Report.com 2012) of mobiles usage reveals that retailer mobile sites are more popular than retailer mobile apps. *The majority of mobile owners used their devices for shopping and this is only going to grow as their penetration continues to rise.* The study's data was derived from the participation of 5,000 mobile users of Android and Apple devices; the Nielsen data revealed that mobile owners of both genders prefer retailers' mobile sites over mobile apps, with men slightly more likely to try retailers' mobile apps than women.

However, the report did indicate that *consumers who use retailers' mobile apps tend to spend more time on them.* But during the height of the holiday shopping season, all five of the top mobile retail websites—Amazon, Best Buy, eBay, Target, and Walmart—experienced traffic and session length increases that helped narrow the gap between mobile app and mobile site loyalists. During the 2011 holiday season, the preeminent five online retailers listed above reached nearly 60 percent of mobile owners.

Develop both a site and app and retail industry analysts believe that, in the long term, mobile apps will be a better friend to mobile shoppers than a retailer's mobile site. For now, the fact remains that a quality app is more difficult and more expensive to build than a site optimized for mobile. While the Nielsen analysis did not speak to any disparity in quality between mobile apps and mobile sites, there remains a clear difference. But as retailer-made mobile apps improve in functionality and utility, mobile sites might eventually see some stiffer competition.

So although the relative value of mobile apps and mobile websites will continue to be debated in the retail space, the resounding popularity of each—regardless of which is preferential to most mobile owners—illustrates why retailers are wise to deploy *both* a mobile app and a mobile site as part of their mobile marketing strategy today. *Retailers need to think of their business as a multi-channel environment that can potentially include mobile, online, and brick-and mortar-stores. Winning with shoppers requires a consistent experience across channels that reinforces the values they represent as a retail brand, whether it be price, service, reviews, selection, style, or other key attributes.*

Google released a study titled "Incremental Clicks Impact of Search Advertising" (<http://static.googleusercontent.com/external_content/untrusted_dlcp/research. google.com/nl//pubs/archive/37161.pdf>; Figure 1.13) that showed the amount of search ad traffic that is incremental to traffic from an advertiser's organic search results. In the study, Google asked the following questions: What happens when search ads are paused? How much does organic traffic make up for the loss in traffic from search ads? Google found that an average 89 percent of paid clicks are essentially lost and not recovered by an increase in organic clicks when a search campaign is paused. This number—what Google calls the Incremental Ad Clicks (IACs)—was consistent across all verticals. In the Google study, they only examined

Scenario	Average IAC
Decrease search ad spend to zero (paused)	85%
Decrease search ad spend, but not to zero	80%
Increase search ad spend, from a zero base	79%
Increase search ad spend, from a non-zero base	78%

Figure 1.13   Google Incremental Ad Clicks (IAC) analytics study results.

cases where ads were completely paused. Google looked at three additional change scenarios and included new cases, giving the study a total of more than 5,300 cases.

For the paused cases, the average IAC of 85 percent was a little lower than the previous value of 89 percent. Google found that there was some volatility in this estimate, month-to-month, driven purely by the mix of advertisers who choose to pause their ads in that month. In the cases where spend was decreased (as opposed to paused), Google found that the ads associated with the spend decrease drive, on average, 80 percent incremental traffic. This means that 80 percent of the traffic from those ads would not be made up for by organic traffic. This value is lower than the 85 percent value in the paused cases, possibly due to advertisers selectively turning down parts of their search advertising that they find less effective.

In cases where an advertiser was already spending on search ads and subsequently increased their ad spend, Google also found that the associated ads drive, on average, 78 percent incremental traffic. In the last scenario, where advertisers were previously not advertising with search ads and then turned on search ads, the incremental traffic was 79 percent. Across the board, the Google findings were consistent: ads drive a very high proportion of incremental traffic—search traffic that is not replaced by navigation from organic listings when the ads are turned off or turned down.

1.22 Social Mobile Site

In its IPO statement, Facebook stated that their site had more than 425 million mobile monthly active users; that is a lot of people walking around staring into tiny black rectangles. Today, more and more content is consumed through mobiles. Google Analytics can provide useful mobile traffic information—the developer can simply scroll along their menu on the left and click on Mobile from within the Audience section; then click Devices; in the main window, the developer will be able to see a list of mobiles that site visitors are using.

Before discussing the best practices for creating a social mobile site, let's run through a list of things not to do. There are several good reasons not to do these things because they will slow down mobiles; and if visitors are on a bad 3G connection or they have a low-signal Wi-Fi connection, that can ruin the site visitor experience for them. What the developer wants is a lean, stripped-down, basic version of the traditional site that conveys the most important messages to mobile site visitors.

Here are some important mobile site development concepts and practices to avoid. Developers should avoid tables; the mobile site should consist of one long vertical scrolling column. They should avoid excessive use of JavaScript or PHP or any other code; instead, they should use HTML5. They should avoid the use of large images that can take forever to load; instead, they should use small images that leave a small footprint. They should avoid attempting to recreate the main site;

instead, a mobile site should contain only the most important content that moving visitors need access to, such as maps, coupons, and directions.

Developers should use a media query that allows them to create different design layouts for different screen sizes. Another option is to use a redirect, which should be able to detect if visitors are using a mobile browser, and if so, they should be sent to the mobile version of the site. This is different from a media query, in that the developer is not using the same content but with a different CSS. The question remains on how to make it work for mobiles that do not support JavaScript: the answer is to use PHP, and here is a code example:

```
function mobileDevice()
{
$type = $_SERVER['HTTP_USER_AGENT'];
if(strpos((string)$type, "Windows Phone") != false ||
strpos((string)$type, "iPhone") !=        false ||
strpos((string)$type, "Android") != false)
return true;
else
return false;
}
if(mobileDevice() == true)
header
```

1.23 Web and Wireless Sites

There is also an option that developers may want to try and that is to put a link on their traditional site that asks visitors if they are on a mobile, and if so, they can touch the link to be sent to the mobile version of a site. In a lot of ways, using a link on the main site to a mobile site may be an optional approach. In a best-case scenario, the query or redirect works exactly as planned; in a worst-case example, visitors see the mobile version while on a desktop and they see the main site while on a mobile—or they cannot access the main site while on a mobile.

Just because the developer is going mobile does not mean that they need to lose out on any potential ad revenue; especially if it is a content site. It used to be that a lot of mobiles could not support JavaScript. That is not necessarily true anymore; Google AdSense™ will soon do away with their ads tailored especially for mobile. The AdSense site (<http://www.google.com/adsense/>) currently states that if a site is intended to be accessed by high-end devices—such as iPhones and Android phones—developers can simply use AdSense to generate their ad code. AdSense for content ad code uses JavaScript to deliver text, image, and rich media ads to visitors using high-end mobiles. With this option, developers are able to use this ad code for a variety of ad formats and on different platforms too, whether that is the 320×50 mobile leaderboard, which is usually used for mobile sites, or the 160×600-wide skyscraper, which is usually used for traditional desktop sites.

When in AdSense, Google provides a 320×50 mobile banner ad size, from which the developer can then choose between Text, Image/Rich Media ads, or both; they also have the option to choose their color scheme and then save and get the code:

```
<script type="text/javascript"><! --
google_ad_client = "ca-pub-XXXXXX-X";
/* Mobile Ad */
google_ad_slot = "XXXXXXXX";
google_ad_width = 320;
google_ad_height = 50;
// -- >
</script>
<script type="text/javascript"
src="http://pagead2.googlesyndication.com/pagead/show_ads.js">
</script>
```

Placement of the ad is important, and Google will lower website search rankings if the content above-the-fold is littered with ads. Google does not go into specifics as to how many ads are too many; but if the developer starts to see their page rank slip, then they might want to reconsider having three-fourths of their above-the-fold content being advertisements. Amazon is now earning $1 billion per month in ad revenue simply by placing highly relevant third-party ads at the bottom of their product pages.

It is recommended that developers either place their ads in a far-right column or at the bottom of their content. As noted earlier, developers should not be coding their mobile site using tables; instead, they should limit themselves to one vertical column of content. They should make sure to provide sufficiently solid content to keep their visitors engaged enough to keep scrolling down to reach the ad. Google Analytics code can also track site visitors; Google has specific SDKs for both iOS and Android that will allow developers to track visitors using low-end mobiles that do not support JavaScript. However, most modern Android and iOS mobiles do support JavaScript. Here is some sample code:

```
</head> tag:
<script type="text/javascript">
var _gaq = _gaq || [];
_gaq.push(['_setAccount', 'UA-XXXXX-X']);
_gaq.push(['_trackPageview']);
(function() {
var ga = document.createElement('script'); ga.type = 'text/
javascript'; ga.async = true;
ga.src = ('https:' == document.location.protocol ? 'https://
ssl' : 'http://www') +
'.google-analytics.com/ga.js';
```

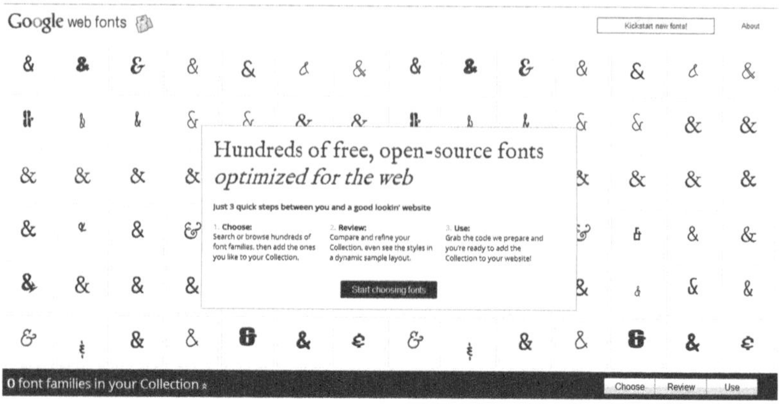

Figure 1.14   Google Web fonts site.

```
var s = document.getElementsByTagName('script')[0];
s.parentNode.insertBefore(ga, s);
})();
</script>
```

While it is very safe for developers to use Arial font for their mobile sites, they should be aware that most Android mobiles will default all fonts back to Droid font, and iOS only supports a limited number of fonts; thus, they should test their fonts to see if they will work before they go into full production mode. One of the best places to find online fonts is Google Web fonts (<http://www.google.com/webfonts/>; Figure 1.14) to add a font. All the developers need to do is add one line of code in their <head> section. Note that when choosing a font, developers should pay attention to the Page Load speed dial; Google will tell them how long the font will take to load for their site.

Another site that provides fonts is called Font Squirrel (<http://www.font squirrel.com/>; Figure 1.15) and they have a @font-face generator. However, they require a ton of code in the CSS; for example,

```
/* Generated by Font Squirrel
(http://www.fontsquirrel.com) on February 28, 2012 04:37:11 PM
America/New_York */ @font-face { font-family:
'TitilliumText22LThin'; src: url('TitilliumText22L001-webfont.
woff')
format('woff'), url('TitilliumText22L001-webfont.
svg#TitilliumText22LThin') format ('svg'); font-weight:
normal; font-style: normal;} @font-face { font-
family:'TitilliumText22LLight'; src: url('TitilliumText22L002-
webfont.woff') format('woff'), url
('TitilliumText22L002-webfont. svg#TitilliumText22LLight')
```

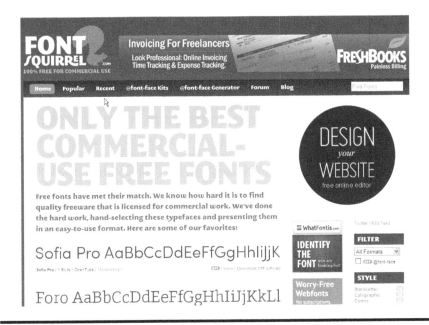

Figure 1.15 The Font Squirrel site.

```
format('svg'); font-weight: normal; font-style: normal;}@
font-face {
font-family: 'TitilliumText22LRegular'; src:
url('TitilliumText22L003-webfont.woff') format('woff'),
url('TitilliumText22L003-webfont.svg#TitilliumText22LRegular')
format('svg'); font-weight: normal; font-style: normal;}@
font-face {font-family: 'TitilliumText22LMedium'; src: url
('TitilliumText22L004-webfont.woff') format('woff'),
url('TitilliumText22L004-webfont.svg#TitilliumText22LMedium')
format('svg'); font-weight: normal; font-style:
normal;} @font-face { font-family: 'TitilliumText22LBold';
src: url('TitilliumText22L005
-webfont.woff') format('woff'), url('TitilliumText22L005-
webfont.svg#TitilliumText22LBold') format('svg'); font-weight:
normal; font-style:normal;} @font-face { font-
family:'TitilliumText22LXBold'; src: url
('TitilliumText22L006-webfont.woff') format('woff'),
url('TitilliumText22L006-webfont.svg#TitilliumText22LXBold')
format('svg'); font-weight: normal; font-style: normal;}
```

Finally, for preview and testing purposes, there is the Adobe tool, discussed previously, called Shadow. This tool allows developers to view their mobile site for testing purposes; developers can download it directly from Adobe (<http:success.

adobe.com/en/na/sem/products/shadow.html?sdid-JRBBP&skwcid=TC|1026867|. adobe%20shadow|S|b|12430492140>).

1.24 Building a Mobile Site with HTML5

All businesses, marketers, and brands understand that mobile is exploding. The number of mobiles in the United States has surpassed the 100-million mark, and mobile Internet usage is continuing to increase at a rapid rate. In fact, a Cisco report (see <http://www.cisco.com/en/US/solutions/collateral/ns341/ns525/ns537/ns705/ns827/white_paper_c11-520862.pdf>) projects that mobiles will exceed the world's population this year. In this new, mobile-driven digital world, businesses are faced with the important challenge of offering customers a mobile experience that is just as comprehensive and easy to use as found on the traditional website. Mobile customers are demanding the ability to do everything they are used to doing on a desktop from the palm of their hands. To accomplish this, marketers and brands should turn to HTML5, the latest HTML coding language that allows developers to deliver a richer and more intuitive user experience within the mobile browser.

What is HTML5? HTML5 allows mobile developers to create more dynamic and engaging Web content for different use cases and user experiences. Optimizing HTML5 features for mobile interfaces enables brands to keep pages light and focus on functionality that takes advantage of the smaller screen size of mobiles. Most new mobile browsers support HTML5; therefore, when building a mobile site, it is very important to make sure that developers support HTML5 functionality for these mobiles. By implementing a well-executed mobile strategy that takes advantage of next-generation HTML5 technology, consumer-facing brands will see an increase in conversion, repeat visits, and overall positive brand awareness from users engaging with their brand via mobiles. Here are some reasons why developing a mobile site and strategy with HTML5 will improve the customer experience.

HTML5 allows the developer to create a rich, app-like user experience in mobile browsers, one that does not require users to actively seek out and download an app through the store or marketplace process. While downloading an app is behavior typically exhibited by brand loyalists, a mobile site allows marketers and brands to reach a much larger audience, including users, much earlier in the customer life cycle. Mobile site interfaces built with HTML5 look more and more like apps every day, and through the ability of HTML5 to recognize gestures, consumers are able to maneuver a site as they would an app. This includes swiping through navigation, and pinching and zooming in on select areas of the mobile site.

Also similar to native apps, HTML5 mobile sites include local storage, which allows site developers to store data within browser memory. This capability results in mobile sites with the ability to store commonly used data within the browser and reduce the number of back-end interactions with the server, such as delivering pages that load much faster than previous-generation mobile sites. One of the most impactful features of HTML5 is that it enables the ability to bring location aware-ness to the browser.

Retailers and local businesses, for example, can simply prompt the mobile site to recognize the user's location upon launch and deliver "find near me" options that identify the closest stores. Additionally, when a user first loads a retailer's mobile site, HTML5 enables it to prompt them with any special time-sensitive deals or offers that are available at nearby retail locations. Expedia.com (<http://www.businessinsider.com/blackboard/expedia>), for example, uses HTML5 to add location-aware context to its mobile site. Upon launching its mobile site, the travel giant leverages the mobiles' internal GPS to offer travelers the ability to search for nearby hotels with same-day vacancies. It is also able to push time-sensitive notifi-cations of special offers to users based on their location.

For retailers, one of the keys to converting visitors into buyers is simplify-ing the shopping cart and checkout process. According to data from Monetate (<http://moneyate.com/infographic/shopping-cart-abandonment-and-tips-to-avoid-it/#axzz1pZj3u8Cp>), 75 percent of all online purchases never occur because people abandon their shopping carts. On mobile, the need to deliver a seamless and easy-to-use checkout process is even more important. HTML5 mobile sites make it easier for brands to implement advanced checkouts that allow users to complete their transaction in one step from whatever page they are on.

To achieve this, retail brands like Tesco allow mobiles to view their carts via overlays on the product and category pages. This means that users can easily call up their shopping cart at any time without having to navigate away from the page that they are on—thereby empowering users with this ability to check out quicker and with many fewer clicks, which can help boost mobiles' conversion rates.

Another big advantage to mobile sites built with HTML5 is their ability to serve high-resolution image galleries. One of the traditional challenges with mobile shopping has been the customer's inability to view detailed images. This is largely alleviated with next-generation HTML5 mobile sites, which allow brands to dis-play multiple, high-quality product images in a gallery format.

HTML5 mobile sites offer users the ability to look through a number of prod-uct views and also double-tap to zoom in on specific images and view products in more detail than previously available. This ability to tap-tap-zoom to get a closer view of an image is especially important on mobile due to their smaller screen size.

Finally, HTML5 makes it much easier for brands to target users with special offers and relevant promotions by providing the ability to scroll through banners on the mobile homepage. Brands can further streamline mobile site navigation and simplify the user experience by integrating expand and collapse menus and pop-up windows on the homepage and category pages—features not previously available but made possible with HTML5. In the mobile-driven world, it is no longer enough to offer a basic optimized mobile site. Instead, brands must develop a strategy that encompasses next-generation features and functionalities to make the customer journey just as comprehensive and easy to use as offered on the traditional website.

Leveraging HTML5 technology is an easy way to improve the consumer's experience with a developer's brand in the mobile browser environment. By developing this mobile site presence with next-generation HTML5 technologies, brands can deliver to users a rich, app-like experience across all major mobile operating systems. The debate between mobile sites versus mobile apps continues to evolve. In fact, recent advancements with HTML5 are making mobile sites ever more powerful, such that the distinction between the mobile Web and apps is blurred. It is important to understand and accept that people will access many channels, and developers should be consistent in their branding and user experience.

1.25 The Importance of Mobile Site Speed

Certain large mobiles, such as tablets, demand high-speed access to mobile sites, and any pages that lag could drive them to other sites. According to a report from Equation Research (<http://www.equationresearch.com/>), most tablet users expect pages accessed via their mobiles to load at a speed similar to their desktop or laptop computer, eMarketer relayed (<http://www.emarketer.com/Article.aspx?R=1008943>). Moreover, 66 percent of consumers said the main issue they encounter with Web access on their mobiles is speed. While marketers cannot fix a slow mobile network, they can be sure that their content aimed at driving traffic from qualified prospects is optimized for mobile access.

Mobile site speed should be a top priority as mobile tablets users are becoming an increasingly large portion of the population. Many are conducting purchase decision research from these handheld devices. Laggard websites could be a turn-off for these ready-to-buy consumers and B2B buyers. Brafton (<http://www.brafton.com/>), a marketing research firm, recently reported that this year's shopping season (2012) was especially strong for tablet sales, as a recovering economy and greater demand for the mobiles increased interest in iPads and various Android-powered devices.

Pew Internet Research (<http://www.pewinternet.org/>) reported that mobile tablet ownership in the United States essentially doubled during the holidays for this year (2012). The research company pegged the number of adults using mobile double to about 19 percent compared with last year's (2011) 10 percent of Americans using these large mobiles. Moreover, tablet sales will likely continue their ascent, as fewer consumers and businesses purchase laptops for their computing. On the other end, many in the market for e-readers have opted for tablets because they get access to the same digital book-reading capability and greater Web access. Additionally, Brafton has reported (<http://www.brafton.com/news/online-purchases-on-tablets-smartphones-becoming-more-popular>) that online purchases via tablets are rising. As such, mobile site speed should become a focus for businesses, marketers, and brands.

Developments in Web design, namely HTML5, have made it easier for companies to create a site fully accessible via mobiles and desktops. However, companies must be sure to address their mobile audience by constantly testing their websites to monitor potential site speed issues. Site speed is an increasingly important element of Web marketing across mobiles, as it has become a ranking signal for search engines. Brafton recently reported (<http://www.brafton.com/news/cutts-page-speed-important-for-seo-but-dont-hide-slow-pages>) that the amount of websites negatively impacted by poor site speed is minimal, but Google continues to urge marketers to monitor this element of their mobile sites. As Google's algorithm shifts to focus more on more mobile user-friendly sites, poor site speed could negatively impact their SEO campaigns.

1.26 Mobile Site Marketing Challenges

The first challenge for mobile site developers is transparency and letting users know what is going on; with desktops, users have a sense that their online behavior is being recognized via such Web mechanisms as beacons and cookie. However, with mobiles, this is not so clear; keep in mind that mobiles are with users continuously in their pockets or purses. Mobiles are the one device that is with users almost all the time, even on their nightstand while they sleep. So users need to know what data their mobiles are capable of communicating.

Okay, so assuming the first issue is addressed, the next issue is having some sort of choice in the tracking of mobiles. Once users know that their mobiles can be tracked by their location, and essentially send the equivalent of computer "cookies" out to marketers and other third parties, they will most likely want the ability to choose whether or not they want that happening. If users like getting great targeted content, and their favorite products and services, they will allow this to take place; if not, they should be given the option to opt out and be able to turn off tracking.

The third and final challenge in the equation is more on the provider end in the mobile arena; although the traditional LinkedIn site (<http://www.linkedin.com/>) has cookies, the LinkedIn app does not, so there is no way to consistently and universally track between the two. The need for a standardized cross-platform approach to mobile privacy is essential to maximize mobile advertising opportunities. The prior mobile industry solution was the UDID (Unique Device ID), a unique string of numbers associated with a device that can let developers track their apps.

1.27 Construction of Mobile Sites Are a Top Priority

Mobile site creation has become a priority for all types of businesses attempting to stay relevant in the fast-changing Internet environment. Projections indicate that in the next two years, over 50 percent of all Internet traffic will come from mobile devices. There is little debate that the meteoric rise in the usage of mobiles to access the Internet is telling of the future. Nielson (<http://blog.nielsen.com/nielsenwire/

consumer/a-store-in-your-pocket-retailer-mobile-websites-beat-apps-among-us-smartphone-owners/>) recently released a study showing dramatic increases in consumer usage of mobiles to shop. Not only are consumers making purchases from their mobiles, but also they are making them through the retailer's mobile site. Consumers are going through every step of the shopping process from product research to purchase, and they are doing it on mobile websites.

If a site is not optimized to accommodate the new trend in retail, mobile shopping, sales will be lost. Mobile sites should ensure that what a user sees is the best possible representation of their brand. Also, buttons and text need to be larger, with clear *Calls to Action*. Functions, especially estimate request forms and shopping carts, need to operate as easily as they do on the traditional desktop website. The easier a mobile website is to use, the more comfortable consumers will be when making a purchase. There are people trying to access sites with mobiles every day.

However, according to Google research (Google Analytics), when people have to wait, they almost certainty abandon the site. Also, when they are waiting for the webpage to load, they are not doing anything—not shopping, not consuming content, and not viewing ads. This has adverse effects on the Internet companies and local companies that are trying to find and enlarge their customer reach. Google found that there is a clear correlation between speed and the success of an online business. More often than not, the problem is because the webpage has not been well designed to load quickly on a smaller device. Reasons could be plenty, such as high-resolution images, data-intensive effects, animations, etc. Owing to that, users simply give up and close down that webpage, which means that companies will lose some potential revenue.

Google is already in the midst of a solution; they are tweaking the settings of their Google Chrome browser for Android—in such a way as to quicken the whole loading process for a webpage. The software will use AI (discussed in detail in Chapter 5) to figure what webpage a user might visit and start loading that page while the user is still typing in the query. This is already available but is in beta testing stages. Google is also figuring a way to reform the age-old Internet protocols that have been accounting for data losses and network connections. One of the revisions is their TCP Fast Open (<http://research.google.com/pubs/pub37517.html>), which means the mobile does not need to be synchronized with the server before transmitting data. If this change is accepted, then the sync will happen instantaneously.

1.28 Mobile Site Metrics

When creating a mobile site, there are some key performance indicators (KPIs) that developers, analysts, and brands need to consider, including the following basic metrics:

1. *Unique versus return users*: Unique impressions equal success and strength of a brand.
2. *Session length*: Measure mobile browsing time daily; the goal is to increase this.
3. *Conversion percentage*: This can help determine the success of a site design strategy.
4. *Mobile information*: This can help identify what dominant devices to target.
5. *User clicks*: Measuring where the users are clicking will guide the site design.

Once all these KPIs are assembled, it is important to cross-reference them—to discover what mobiles have longer sessions, have a higher conversion and, most importantly, generate the most revenue. Mobile site modeling also covers the adoption, engagement, and effectiveness of mobile marketing campaigns coupled with app analytics to gain insight into the number of downloads and how users are interacting with different parts of the mobile site.

This data helps mobile analysts, marketers, and brands quantify the ROI of ad campaigns, including mobile search buys, click-to-call, SMS volumes, QR code swipes, and app ads. Among the mobile site metrics analysts, developers, and marketers may also want to include are the number of page views, bounce rates, and click behavior; such data is useful in optimizing mobile sites and app design.

But mobile modeling goes beyond reporting on these basic KPIs. To take it to a higher level, mobile site analytics requires the use of data mining, business intelligence, and—in the end—AI modeling technology using clustering, text, and classification software. The essence of mobile site analytics is to offer the content, products, and services—most aligned to mobiles' desires, values, wants, and needs—at the microsecond that they occur.

Yes, *reacting* to SMS messaging, QR codes, mobile apps, and mobile site behaviors at the instant they take place is important, but *anticipating* these mobile behaviors requires much more than merely aggregating log file analyses and KPI reports. Marketers, enterprises, agencies, and brands looking to truly leverage mobile sites and analytics are faced with some unique challenges and opportunities. On the one hand, data collection can be more difficult—for example, some devices do not accept cookies or use JavaScript, which is the most common way to track Web data; but because of the triangulation of mobiles via other mechanisms such as GPS and Wi-Fi can be used for mobile analytics.

Mobile analytics provide an opportunity to truly segment mobiles based on attributes such as type of device, network access, location, and interests. This creates multiple opportunities for marketers and brands; it is becoming more important every single day as mobiles edge out stationary devices as the way consumers access the Internet in search of products and services.

Deloitte (<http://www.marketingcharts.com/direct/mobile-devices-to-overtake-pcs-this-year-15836/>) predicts that mobiles will reach almost half of the total computing market size by the end of this year. InternetRetailer (<http://www.

internetretailer.com/2011/03/03/20-mobile-phone-owners-use-mobile-web-every-day>) found that one in five U.S. mobile device owners use their mobiles to access the Internet every day. The bottom line is that if companies do not embrace mobile sites and apps, they may be at a huge disadvantage in the years ahead because this is where consumer data is being created for behavioral analytics. Although many mobile analytics vendors exist, the best way to begin may be to start with Google's free mobile site Web analytics solution (<https://developers.google.com/analytics/devguides/collection/>).

Brands, marketers, and developers can also try using another solution that is more specialized; for example, PercentMobile (<http://about.delivr.com/>) is an analytics vendor that focuses on mobile site reporting. This company simplifies deployment by delivering their mobile analytics software as an on-demand, hosted service capable of reporting on large amounts of mobile site traffic. There is no need for software, dedicated hardware, or the purchase of an appliance. A mobile site simply installs their platform-specific tracking code into their templates to start tracking in minutes.

When creating their mobile sites, enterprises, developers, marketers, and brands should address presentation technologies that support multiple devices, networks, and operating systems. They also should give consideration to incorporating mobile Web analytics as the core to better support and engage visitors anytime, anywhere. We now move to another important platform for aggregating mobile data: apps.

Chapter 2

Mobile Apps

2.1 Why Apps?

Chapter 1 discussed the importance of mobile sites; however, a site is often browsed momentarily for anything interesting and then the browser is closed. In contrast, an app is installed, used, and remains on the mobile until the owner chooses to remove it. As such, an app is likely to be opened again and again. If the content and information about products and services are valuable, the user keeps the app; however, the stream of relevant content is vital, such as a price comparison feature that is dynamic. This means more repeat traffic, increased consumer loyalty, and a good way for brands and marketers to drive engagement with consumers. Because of current consumer behavioral trends, having *both* a mobile site and an app is highly recommended to brands and marketers.

Today, the browser is moving and, as such, apps can be used for targeting mobiles; users are demanding immediate access to relevant mobile data in this new channel. They want Web-based email, news, shopping, weather, games, sports, and entertainment content, which mobile analytics can deliver by segmenting and classifying apps downloads and behaviors after installations. Due to this shift to mobile—consumers are *always-on and always-connected*—location-aware mobiles represent a strategic new channel that retailers and brands must recognize as fundamentally different from other marketing conduits.

Brands, enterprises, marketers, and retailers should take a strategic approach to deliver a unique mobile shopping and in-store experiences. Apps can be used to target ads that are uniquely relevant to consumers based on their previous behaviors and their physical location. A retailer, for example, can use apps to market to mobiles to enhance customer engagement and loyalty. The mobiles are a convenient way for customers to gather information about products and services, and even

61

conduct transactions. Consumers can research prices and do comparison shopping by scanning barcodes using their mobiles. The key is to provide convenient features in order to retain consumers' interest and loyalty via the mobiles in their pockets or purses.

There are a number of tools designed to help non-developers create apps, or let developers create apps in the programming language with which they are most familiar and convert them to the appropriate format for Apple or Android devices, the dominant operating systems. For programmers, the options include appcelerator® (<http://www.appcelerator.com/>), MOTHERAPP® (<http://www.motherapp.com/>), and Adobe PhoneGap™ (<http://phonegap.com/>), while for non-developers, there is AppsGeyser (<http://www.appsgeyser.com/>), RunRev (<http://www.runrev.com/>), AppMakr™ (<http://www.appmakr.com/>), GENWI™ (<http://genwi.com/>), SaasMob™ (<http://saasmob.com/>), MobBase™ (<http://www.mobbase.com/>), and many more. It is worth noting that recent research at the sites listed above has found that consumers interacting with apps show more brand favorability and purchase intent. The study also found that apps that were more informational and generic in nature, such as providing product reviews, deals information, coupons, or tips, were more effective in engaging users, as compared to games or entertainment apps.

Having the ability for consumers to engage with retailers encourages *"situation shopping,"* in which they can mix and match clothes through a store-specific app. For example, for a clothing retailer this can create centric-search features and a virtual closet—incorporating different colors and styles—where mobile shoppers can try out different outfits or items via a retailer's app. Apps increase consumer browsing time and loyalty and for the retailer or brand, improved sales.

Retailer-specific apps can be developed for different types of products, such as those for electronic gadgets, auto enthusiasts, do-it yourself hardware, runners, gardeners, golfers, etc.—the potential is endless. These retailers can incorporate store-specific mobile apps, which can include scanning capabilities for price comparison abilities not only in their store, but also across all retail sites on the Internet. However, the most important goal is not to get a marketing message across, but instead the developer should think about what consumers actually need. Serve the customer, because if customers cannot see any value in the app, they will either never download it or promptly forget about it once they do download it.

2.2 Google Mobile AdWords App

To assist app developers, Google has introduced four new tools—a new extension for their AdWords platform and additional information to the click-to-download format—with the ability to see Google Play (<https://play.google.com/store?hl=en>) stats in AdWords, and Custom Search Ads for mobile apps (Figure 2.1). The Mobile App extension is currently only available for mobiles and not tablets, which will

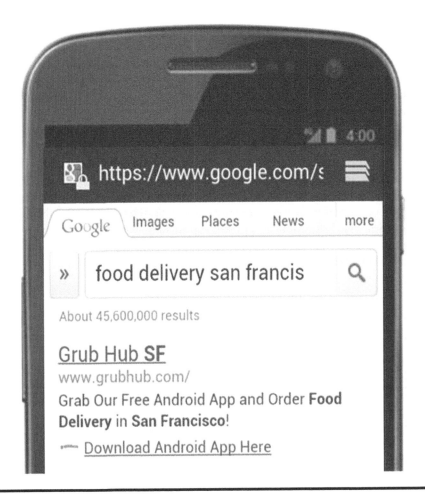

Figure 2.1   Free Android™ app.

help advertisers raise awareness of the fact that they have an app. When users are searching on their mobiles for a brand name or product category, they will see relevant ads displayed, and advertisers that have apps can append a link to their app below the ad description. If a user is searching for Wells Fargo® or Walgreens®, for example, the ad would allow users to see that they could get information via the brand's app, rather than going to the mobile site.

AdWords with the Mobile App extension will have two links: one from the headline to the regular landing page, and a second one on the app that will lead to Google Play or the Apple App Store. To set up a Mobile App extension, advertisers need a mobile app that is live within one of the two marketplaces. They will also need the Package Name if it is Android, or App ID if it is iOS, along with the URL of the page within the app store where users can install or learn more about the app.

The Mobile App extensions are available within the Ad Extensions tab within the Campaigns tab in AdWords. Google has also added additional information to click-to-download ads. When users search and see the click-to-download ad, they will be able to see image previews, a description of the app, and, when applicable, pricing and rating information. The additional information will be drawn automatically from Google Play and the Apple App Store.

If they are marketing an Android app, advertisers will be able to track downloads from Google Play within their AdWords account; they will appear as AdWords conversions. Advertisers can set this up via the Conversions tab under the Tools and Analysis tab.

Finally, Google has introduced an ad type for app developers, publishers, retailers, and brands that want to incorporate Custom Search within their tablet apps. Custom Search Ads for Mobile Apps will appear along with organic search results when a user performs a search within the app. Google shares revenue generated with the developer, publisher, retailer, or brand.

2.3 HTML5

Increasingly, apps are being built using HTML5, which is a language for structuring and presenting content for the Internet. HTML5 is also an attempt to define a single markup language that can be written in either HTML or XHTML syntax and is the dominant technology used for constructing apps. The language includes detailed processing models to encourage more interoperable implementations; it extends, improves, and introduces markup and application programming interfaces (APIs) for constructing complex mobile sites and apps.

HTML5 is the ideal language for constructing cross-platform mobile apps because many features of the language have been built with consideration for being able to run on low-powered mobile devices. One important feature of HTML5 is that it allows developers to build new kinds of features into mobile sites so that they behave on mobiles like apps. HTML5 has soared to new heights under the aegis of tech titans such as Google, Apple, Microsoft, and others.

The problem is particularly acute in games, where Apple mobiles expect to be able to interact with Android devices. For example, GameClosure (<http://game-closure.com/>), a start-up, offers a JavaScript game SDK (software development kit) and HTML5, which runs on mobile, tablet, and browser devices, and allows client/server code-sharing. Using only JavaScript on the client and server, developers can build games to enable Android, iPhone, and Facebook users to compete and collaborate in real-time.

At its core, HTML5 is a set of standards that lets browsers understand animation, videos, graphics, and other multimedia content without the need to download a plug-in like Adobe Flash, which is how most traditional website videos and graphics are displayed today. Many technologists—including Steve Jobs, while alive—have criticized Flash for being slow, buggy, and a drain on the limited battery life of mobiles. HTML5 is rapidly making its way into all mobile browsers. Game Closure HTML5 code allows for write-once, publish-anywhere tools for all types of mobiles. Zynga (<http://zynga.com/>) ElectronicArts, Amazon, and Pandora are using HTML5 in their sites and apps.

The ability to deploy on a wide range of mobiles is the most attractive quality of HTML5. The sheer number of devices, models, and operating systems on the market, and rapid adoption by consumers, means that enterprises, marketers, and developers can no longer pick and choose which device they support. Companies and brands are being forced to use HTML5 to create apps that stretch across multiple platforms. HTML5 will not replace native apps as consumers continue to demand a combination of the two; instead, HTML5 will act as an enhancement to existing mobile strategies.

Mobile usage will undoubtedly continue to rise; Gartner (Gartner.com) has projected that the emergence of media tablets and the more than 5 billion mobiles in use worldwide will grow to more than 6.7 billion connections by 2015. Many, if not all, of those mobiles will include HTML5-enabled browsers and will provide an alternative to native applications.

In fact, it is predicted that half of the apps that would be written in platform-specific native languages today will be written exclusively in HTML5 by 2015. As mobiles continue to gain momentum, the integration of a brand strategy has become an essential component for every enterprise and brand. The question for each is: How many of those mobiles can a company reach with their current mobile strategy? There are many architectural styles to choose from: thick, rich, and thin clients; streaming; and messaging. And there are a growing number of channels to deliver them: Apple, Android, BlackBerry, Windows, Palm®, and Symbian (Figure 2.2).

Whether a brand is looking to connect with customers on the move, boost employee productivity, or even strengthen relationships with their partners, developing for mobility requires an action plan for the most optimal application on many channels. While CIOs expect to support their apps on three operating systems per employee, gone are the days where apps are deployed to a single device owned and managed by the enterprise. The mobile strategy must include personal devices whose apps are secure for business.

No matter what approach a developer chooses—mobile web, native, wrapper, or hybrid—it is critical that their choice of technology enables a cohesive design canvas, simple deployment, sufficient mobile API use, and a supportable, long-term, ongoing management model. An app developer's approach should allow for a cost-effective change that is flexible, allowed at any stage of development, and ultimately giving them the power of choice (Figure 2.3).

The question for a brand, developer, business, or enterprise is whether to develop Web apps, native apps, or some combination thereof. Answers to that question are anything but simple. Who is your target audience? What is the purpose of the

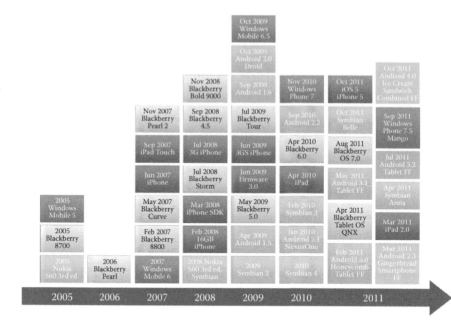

Figure 2.2 The rapid evolution of mobile operating systems.

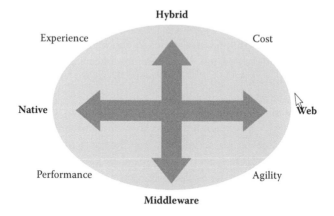

Figure 2.3 The app development options.

app? There are a series of diverse questions that must be answered before jumping head-first into development. Here are the options and the strengths/weaknesses of each choice: Native gives the best user experience, performance, and access to device APIs; camera, contacts, phone state etc. The problem with native is that it is difficult and fragmented between the four major mobile platforms – Apple, Android, Windows Phone, BlackBerry – because each uses a different code base.

The Forrester research firm (Forrester.com) says that enterprise development shops that plan on writing distinct native apps for each platform should plan on a budget 150% to 210% higher than what might be reasonably expected.

The other option that is increasing in popularity is to build an app with HTML5 *and* JavaScript. This is where the mobile browser is ubiquitous; new HTML5 tags have evolved for audio and video, and device access to APIs has dramatically increased the capability of their functionality. Yet, JavaScript is still not as fast as native code; the user experience is often left wanting or not responsive in their design in which developers or enterprises may not want to deal with.

The solution for the time may be to code HTML5 and JavaScript with a wrapper that gives it native capabilities. This is how Facebook does it in a variety of apps such as PhoneGap. There are various Integrated Developer Environments (IDEs) that are used to develop apps on client- and server-side components, a common practice with Oracle or SAP shops, with other more and more flexible and rapid options evolving.

2.4 In Your Face

Almost 40 percent of Facebook users are accessing the site via mobiles according to Facebook.com/Enders Analysis. The monthly active users of the social network site are via mobile apps, which exceeded 300 million in 2012. That is 37.5 percent of the 800 million total monthly active users of Facebook, making it one of the most mobile-centric online services out there (Figure 2.4).

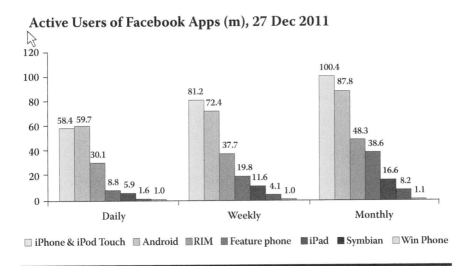

Figure 2.4   The Facebook traffic by mobiles. (From Facebook, Enders Analysis.)

Half of all active Apple and Android users—approximately 225 million mobiles and 70 percent of all BlackBerry users—have installed the Facebook app according to Google Analytics; in other words, almost three out of four mobile users used either the Facebook or other social networking apps to access the site. This is an important statistic that developers, marketers, and brands need to consider in the development of their app projects.

Facebook recently saw the number of monthly active users accessing its mobile apps pass the 300-million mark. With Apple and Android, apps account for more than two-thirds of mobiles on Facebook. The rise in these numbers is likely attributable to the "Christmas rush" during recent years, with Android and Apple mobiles among the most coveted gifts this year (2012). These are dominated by the two platforms that have traction; Android has now passed Apple in daily active users (DAUs), although Apple has passed the 100-million monthly active user (MAU) figure. However, Windows Phone and Research In Motion (RIM) BlackBerry face a different plight.

Windows Phone remains quite insignificant, although that may change as Nokia's efforts come fully on-stream. Meanwhile, around 70 percent of RIM's 70 million active users have installed the Facebook app. That is a high penetration rate on what is supposed to be a corporate product, pointing to RIM's strength in messaging, but also to the way that the mix is shifting away from enterprise customers and toward emerging consumer markets.

Yet perhaps the most interesting analysis is where Facebook is beginning to look at Apple and Android applications as one platform. Apple and Android apps are on the way to being one platform, with Facebook moving them more and more toward being wrappers for a common HTML5 experience. Facebook will treat that user base as less of a mobile extension to the desktop experience and more as the core product, starting with advertising. Flurry Analytics (<http://www.flurry.com/flurry-analytics.html>), a big data analytics company that measures mobile app trends, released its own analysis of Android and Apple activations over the last Christmas holiday. Activations jumped to more than 6.8 million, a 353 percent increase over the baseline of previous years.

2.5 Brand Apps

As more and more people turn to mobiles for everything, an increasingly large number of brands are developing mobile apps to keep up. For example, Krispy Kreme® developed a mobile app that lets users track the nearest Krispy Kreme location to them, no matter where they are. MobileCommerceDaily reports (<http://www.mobilecommercedaily.com/2011/12/28/krispy-kreme-uses-mobile-to-bolster-foot-traffic>) that Krispy Kreme developed the app after learning that 30 percent of their website and social media traffic was coming from mobiles. The Krispy Kreme app uses GPS technology to track the closest store, highlights their new and seasonal products, and features a map and a function that allows a user to call the store with one click.

Chili's® is another company getting in on the mobile app action. They recently developed a convenient app for Apple and Android mobiles that lets the consumer find the nearest location to them, browse their menu, and even place an order. Once a customer places an order through the Chili's mobile app, they can use the app to get directions to go pick up their food.

McDonald's® is also looking to boost sales through geotargeted mobile ads viewable within the Pandora app. When consumers are using the Pandora mobile app, they will see an ad for a McCafe™ or hot chocolate for $1.99. If users click on the ad, they will be directed to a landing page that tells them more about the beverages and lets them find the McDonald's located closest to them. There are countless brands, big and small, developing mobile apps to boost both in-store and online traffic. Local businesses should consider developing their own brand app as more and more consumers are using them to search and shop from their pocket and purses; this will allow them to not have to talk to someone face to face or have to call them on the phone to place an order.

2.6 Apps Metrics and Trends

Upon installation and registration, apps capture important user and device attributes for segmentation marketing. They can capture age, gender, and locations; they also can report on the type of device, model, and operating system. For the brand, analyst, developer, and marketer, apps can also measure and report on which features are the most popular, which demographic groups use the app the longest, and which devices share their apps with others.

It is important for marketers and brands to know what trends are evolving via hundreds of thousands of apps; this can be a daunting task, and market research firms such as PositionApp™ (<http://www.positionapp.com/>) can be enlisted to monitor them. PositionApp can be used by developers and marketers to track the success of their apps, and to measure their impact on sales and revenue in markets all over the world; they provide up-to-date, country-by-country position performance stats, which can be browsed by: country, genre, position change, app name, free, and fees paid on a daily, weekly and monthly basis.

Yet another app metric reporting firm is Appolicious™ (<http://www.appolicious.com/>). They offer a directory of hundreds of thousands of apps; they combine social networking, journalism, and technology to report on apps metrics. Brands, developers, and marketers can register their app in this vast directory and then track its performance around the world. The Appolicious directory includes such categories as ArcadeGames, Baseball, BoardGames, Chicago, Children'sBooks, Education, Golf, London, Movies, NewYorkCity, Olympics, Recipes, SanFrancisco, Travel, etc. They also have such categories broken down by BookLovers, CasualGamers,

App Leaderboard

Rank By: MAU | DAU | DAU/MAU

1.	Yahoo! Social Bar	39,600,000
2.	CityVille	39,500,000
3.	Socialcam	39,500,000
4.	Static HTML: iframe tabs	39,300,000
5.	Texas HoldEm Poker	36,000,000
6.	Viddy	35,500,000
7.	Draw Something	33,700,000
8.	MyCalendar - Birthdays	29,200,000
9.	Bing	29,000,000
10.	Scribd	24,200,000
11.	Dailymotion	24,000,000
12.	FarmVille	23,900,000
13.	CastleVille	23,500,000
14.	Angry Birds	23,200,000
15.	Hidden Chronicles	23,200,000

Developer Leaderboard

Rank By: MAU | DAU | DAU/MAU

1.	Zynga	266,429,278
2.	Microsoft	64,500,000
3.	Thunderpenny	61,810,000
4.	Yahoo!	54,604,345
5.	Woobox	46,070,000
6.	King.com	45,680,000
7.	wooga	44,937,000
8.	Electronic Arts	44,480,108
9.	Socialcam	39,500,000
10.	Viddy, Inc.	35,500,000
11.	MyCalendar	29,200,000
12.	Telaxo	25,645,060
13.	Scribd Inc.	24,200,000
14.	Playdom	24,092,683
15.	Dailymotion	24,040,000

Figure 2.5 The AppData Score Card. (From http://www.appdata.com/.)

Commuters, FitnessFreaks, MusicLovers, NewsHounds, Parents, Shopaholics, Shutterbugs, TVJunkies, etc.

Still another app marketing firm dedicated to monitoring mobile apps is AppData™ (<http://www.appdata.com/>); they provide metrics and the tracking of *apps, social platforms, the social networks,* and *virtual goods.* AppData is an independent apps traffic tracking service of Inside Network® (<http://www.insidenetwork.com/>), a company dedicated to providing business information and market research to the Facebook platform and social gaming ecosystem. AppData is intended for use by brands, developers, investors, marketers, and analysts interested in tracking app traffic on the Facebook platform (Figure 2.5).

2.7 App Exchanges

Mobclix™ (<http://www.mobclix.com/>) is a mobile ad exchange that matches dozens of ad networks with thousands of apps-seeking advertisers. The company collects mobile device IDs, encodes them to mask them, and assigns them to interest categories based on what apps people download and how much time they spend using an app, among other factors. By tracking a mobile phone's location, Mobclix can calibrate the location of the device, which it then matches to demographic data from the Nielsen Company. Mobclix can place a user into more than 150 consumer segments that it offers to advertisers—from "die-hard gamers" to "soccer moms"—by segmenting apps rather than individuals. Mobclix is accurate in targeting mobiles without identifying individuals.

Mobclix provides a supply-side platform for real-time bidding (RTB) of ad inventory. Demand-side platforms (DSPs) and RTB ad networks can bid for individual ad impressions in real-time, giving advertisers and agencies dramatically better targeting and increased performance of their apps. The RTB system gives publishers complete control over every impression so they can maximize revenue, monetize relevant audiences beyond typical strategies, and achieve improved lifts of their ad inventories up to 40 to 85 percent.

Market researcher Gartner Inc. estimates that worldwide apps sales this year will total $6.7 billion (Gartner.com). Many developers offer apps for free, hoping to profit by selling advertising inside the app itself. Users are willing to tolerate ads in apps to get something for free. Ad sales on mobiles is currently less than 5 percent of the $23 billion in annual online advertising, but spending on wireless ads is growing faster than the market overall.

Many ad networks offer software development kits (SDKs), such as Millennial Media® (<http://www.millennialmedia.com/>) for capturing age, gender, income, and ethnicity, to assist advertisers, publishers, developers, and marketers in providing more relevant mobile ads via customized apps. Millennial Media's MYDAS technology engine (<http://www.millennialmedia.com/data-technology/mydas-technology/>) leverages unrefined user data and aggregates it into actionable audience profiles based on key behavior, location, and content trends. These profiles, when coupled with multiple layers of mobiles data, create specific and targetable audiences for advertisers via apps.

Google is the biggest data aggregator for monetizing apps via its AdMob, AdSense, Analytics, and DoubleClick units to let marketers, developers, and advertisers target mobiles by location, type of device, age, and other demographics. Apple also operates its iAd network for marketing on mobiles; in addition, Apple targets its ads to mobiles based largely on what it knows about them through its App Store and iTunes music and video store. The targeted marketing to these mobiles can be based on the types of songs, videos, games, and apps a person has purchased from Apple over time.

2.8 App Developer Alliance

Mobile application developers are set to gain an industry association that would promote their interests, enabling collaboration and product testing while offering education, cloud hosting, and governmental lobbying on their behalf. While Web developers also would be welcome, the new app developer alliance is initially geared toward mobile development for platforms including Apple iOS, Google Android, and RIM BlackBerry.

The new mobile app industry alliance is looking to recruit thousands of developers. There is an interesting space in the application developer community that is not really organized and could use these new service offerings. Key services included as part of the alliance include a collaboration network, via an online database with product-testing facilities offering access to multiple platforms and tools, discounted and free tutorials on trends and technologies, as well as structured training and certification programs, discounted hosting and cloud services via Rackspace® (<http://www.rackspace.com/>).

Lobbying for government policies to assist app developers is expected to be part of the initiative. There is developer interest in an industry-wide privacy policy, as well as IP policies pertaining to patents and copyrights. Mobile broadband policies are also part of the alliance's efforts. Expected backers of the alliance include Google and RIM, while Apple and Microsoft are not participating. The alliance will look to generate revenues via sponsors and, over time, membership fees from app developers.

2.9 Entertainment Apps

Real-time entertainment traffic dominates the Internet now, and over half of it happens on mobiles, this according to a new report by research company Sandvine®. The report (<http://www.sandvine.com/downloads/documents/10-26-2011_phenomena/Sandvine%20Global%20Internet%20Phenomena%20Spotlight%20-%20North%20America%20-%20Fixed%20Access%20-%20Fall%202011.pdf>) states that, by volume, 55 percent of real-time entertainment traffic is destined for mobiles. Those statistics show just how far consumers have come in the post-PC era; of the non-computer traffic, much of it comes from Netflix, Facebook, and YouTube, mostly on mobiles, as such apps are a good vehicle for data aggregation for mobile analytics. Real-time entertainment is defined in the report as apps that allow "on-demand" entertainment, that is, consumed, viewed, or heard as it arrives; examples include Netflix, Hulu, YouTube, Spotify® (<http://www.spotify.com/>), Rdio™ (<http://www.rdio.com/>), Pandora, and Slingbox® (<http://www.slingbox.com/>).

Looking at overall entertainment traffic, which includes both stationary and mobile devices, real-time entertainment accounts for 60 percent of peak downstream Internet traffic in North America, with a steady increase in this figure over the past few years. Moreover, screen size has a direct correlation with data usage. Looking specifically at mobiles, Sandvine reports that real-time entertainment generates 30.8 percent of peak demand with Internet browsing being next, at 27.3 percent, while social networking is only 20.0 percent (<http://www.readwriteweb.com/archives/55_of_real-time_entertainment_is_consumed_on_tv_mobile_tablet.

php>). Most of the latter comes from Facebook, which represents 19.3 percent of peak mobile traffic, with YouTube getting 18.2 percent.

Pandora and Twitter have 60 percent and 55 percent, respectively, of their traffic going to mobiles.

These statistics show that mobiles are not only having a big impact on consumption of real-time entertainment, but they are now the *primary way to consume such content*. Thus, it is no surprise that Apple recently filed a patent application for a system for placing and pricing ads based on a person's Web or search history and the content of their media library.

The Apple patent application also lists another possible way of targeting ads "via the content of a friend's media library." The patent states that Apple would tap "known connections on one or more social-networking websites" or publicly available information or private databases describing purchasing decisions, brand preferences, and other data. Apple recently introduced a social-networking service within iTunes, called Ping, that lets users share music preferences with friends.

2.10 News Apps

A mobile app known as Pulse (<https://www.pulse.me/>) with 13 million users, and more than 2 million new ones joining each month, is a popular news-reader that adds local content and deals. The company could be the next significant media outlet providing international and neighborhood news; brands and developers should take note of this app's strategy (Figure 2.6).

Pulse is one of few default apps installed on the Amazon Kindle Fire, and the app reader has partnerships with over 250 publishers, such as Fox News and Bloomberg Businessweek. Pulse is a start-up developed by AlphonsoLabs, which is moving to integrate neighborhood news and local information from publications including AOL Patch (<http://www.patch.com/>) (Figure 2.7).

There are now a multitude of mobile apps that aggregate news from various publications and present articles and blog posts in a reader-friendly format, including readers from Google and Yahoo!, and a sleek and popular app from Flipboard (<http://flipboard.com/>), which mimics the feel of turning a printed page (Figure 2.8).

Other local news feeders include CBSLocal and community sites such as BoldItalics in San Francisco with unique local intel and backstories about the city by the bay.

Flipboard is another popular news and magazine app with a striking interface. But Pulse is moving to feature daily deals from local merchants, courtesy of Groupon® (<http://www.groupon.com/>), LivingSocial® (<http://www.livingsocial.dom>), and GiltGroupe (http://www.gilt.com/; Figure 2.9).

Pulse will display each local news story or deal as a colorful title—just as articles are displayed by major publications; they will arrange them in horizontal rows

Figure 2.6   Pulse news streaming app.

Figure 2.7   AOL Patch local news portal.

Figure 2.8   Flipboard is a beautiful news magazine app.

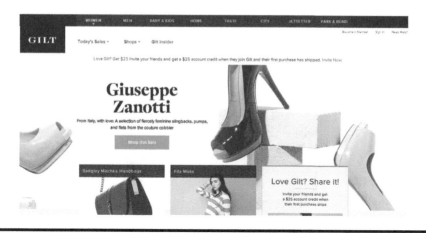

Figure 2.9   Gilt for daily deals.

for easy navigation. These types of news apps will aid local news sites that participate and will benefit from extra traffic and exposure to millions of new users. They provide a way to reach a new and growing audience of consumers. News apps can consider allowing users to sign up for subscription content and taking a cut of each sale. These types of apps are better than relying on intrusive ads that dominate publishing company websites. One option for developers, brands, and marketers is to deliver branded-content ads, stories, and videos that get slotted into the same flow as news content.

Spending on ads placed in mobile apps will rise to 21.2 billion in 2016, according to ad consultancy BorrellAssociates (<http://topipadfinanceapps.com/the-pulse-app-goes-local/>). The new push into local content, for example, will also give news apps like Pulse a better idea of a user's location and help advertisers geo-target mobiles and their owners. Pulse, as with Flipboard, is expanding into international markets in support of multiple languages. The Flipboard user base more than doubled after releasing its first Apple app. Developers need to be aware of the mobiles they support and their local or international markets for their brand—in short, journalism is being reinvented.

2.11 An App To-Do List

Mobile analytics and marketing means focusing on consumer choices that lead to actionable insight and measurable action. It consequently leads to the monetization of mobile behaviors. The use of clustering, text, classification, and streaming software tools enables brands, developers, analysts, and marketers to calibrate how mobiles behave and how they can profit and respond with real-time relevancy in an accurate and effective way. The most important issue of apps is the location, aggregation, and use of the right data with the correlating consumer needs and preferences.

Marketers do not want to offer mobiles products or services they do not want, or content that is not relevant to them; this is not only wasteful, but also intrusive. Instead, apps should be constructed and used—based on historical behavior models—in order to provide the right product to the right consumer at the right time and place. Mobiles' behaviors are the most valuable assets marketers and brands have, and they need to protect them and not share them with others. The following provides some of the best practices on apps construction, mobile analytics, and strategic deployment:

1. Design the app to make the consumer experience so unique it will ensure their loyalty for life—apps should "anticipate" the mobiles' preferences, desires, and activities. To accomplish this, the clustering and modeling of mobile behaviors must be continuously refined by brands, developers, and marketers.
2. Ensure that an enterprise's IT systems and the behaviors they capture for mobile analyses are aligned with the business goals of the company and its marketing efforts. For example, if a marketing campaign is being planned via apps, the IT department must be prepared to construct or contract for the creation of a mobile site to complement it. The marketer must also be prepared to target, at the very least, Android and Apple mobiles.
3. Incrementally measure the results of the unsupervised and supervised models to optimize their performance. The mobile marketer must understand that this is an ongoing process and not a one-time project. If streaming analytical software is being used, the developer and marketer must monitor the results of the inductive or deductive rules to ensure optimum performance, yield, sales, revenue, and relevancy.
4. The app developer and mobile marketer must recognize that analytical systems must be flexible and adaptive to change within a rapidly evolving business environment—model often, and if possible continuously—developers should measure *everything*. Listen to what mobiles are saying about a brand or a company. As previously mentioned, there are a number of firms and techniques that can be enlisted to accomplish this.
5. Leverage existing IT legacy systems with external analytic services, such as psychographics networks, ad exchanges, recommendation engines, and social media. The app developer and marketer need to conduct a comprehensive data audit of the internal and external data sources in order to create a framework for strategically developing an app to ensure mobiles are being served with relevant information about their content, products, and services.
6. Protect consumers' privacy and respect their desire; state and share clearly why analytics and modeling is being performed: to ensure consumer relevancy and improve customer service. The marketer needs to ensure that anonymous

techniques and technologies are being used properly, such as mobile cookies and digital fingerprinting, which target mobiles rather than human consumers.

7. Recognize that every company is unique, so its analytical strategy, components, architecture, and design will be driven by its industry and marketplace, as well as by the type of product or service it offers to consumer mobiles. The responsibility for ensuring the alignment of these factors rests with the developer and marketer and how the app is constructed, designed, and deployed.

2.12 "Picture-This" Apps

The iTune depository offers more than half a billion apps, with the camera category having split into dozens of subbranches. There are photo editing apps such as Camera+ (<http://campl.us/>), which enables users to use their Apple mobiles to shoot the best photos they possibly can. Packed with several useful features, this photo app is the most popular photo-editing app; it sells for 99¢ and has made its creators over $5 million. Among other photo-editing apps are iPhoto®, a digital photograph manipulation software application developed by AppleInc. and originally released with every Macintosh personal computer as part of the iLife® suite of digital media management applications. First released in 2002, iPhoto can import, organize, edit, print, and share digital photos and now in an app for all Apple mobiles.

Other photo editing apps include AntiCrop™ (<http://adva-soft.com/#>) for cropping, or Juxtaposer™ (http://www.pocketpixels.com/Juxtaposer.html>) and ColorSplash™ (<http://www.pocketpixels.com/ColorSplash.html>) for silhouetting and design. Then there is Microsoft Photosynth® (<http://photosynth.net/>) for creating 3D images using Microsoft technology on Bing. Hit the right magic mix of photo features and a camera app might be acquired by Facebook just like Instagram (<http://instagram.com/>) for $1 billion! (see Figure 2.10).

Other photo-editing mobile apps for professionals include DynamicLight, CameraGenius, Photogene2, IrisPhotoSuite, 360Panorama, Autostich, and Pano. There are also morphing photo apps such as Fatify, Younicorn, TinyPlanet, and Splice-O-Matic for altering pictures on Apple and Android mobiles. There are also photo-booth apps such as Popbooth, Incredibooth and PocketBooth—as well as apps for adding sound to photos via StoryMark, Tap2Cap, and PhotoSpeak—and sideshow apps such as Animoto, SlideShowBuilder, and PhotoSlideshowDirector. Then there are filter mobile apps such as Hipstamatic (<http://hipstamatic.com/the_app.html>)—where true hipsters used this retro tool—which makes over $10 million a year (Figure 2.11). Other app filters for graffiti, artsy, and vintage photo features include CatEffects by Daniel Cota (<http://cateffects.com/>), HelloKittyCamera by AITIA Corporation (<http://itunes.apple.com/us/app/hello-kitty-camera/id317065148?mt=8>), Superimpose by Pankaj Goswami (<http://itunes.apple.com/us/app/superimpose/id435913585?mt=8>), 8 mm Camera by Nexvio Inc. (<http://itunes.apple.com/us/

Figure 2.10 Social photo app Instagram.

Figure 2.11 Hipstamatic, the cool $1.99 filter app.

app/8mm-vintage-camera/id406541444?mt=8>), PencilSketch by H. Rock Liao (<http://itunes.apple.com/us/app/pencil-sketch-hd/id421778766?mt=8>), Pixel Pix by XOXCO, Inc. (<http://itunes.apple.com/us/app/pixel-pix/id501717166?mt=8>), Tiltshift Generator by Art & Mobile (<http://itunes.apple.com/us/app/tiltshift-generator-fake-miniature/id327716311?mt=8>), CaricatureMe by Miinu (<http://itunes. apple.com/us/app/caricature-me/id495103673?mt=8>), HopePosterPhotoFilter by 3DTOPO Inc. (<http://itunes.apple.com/us/app/hope-poster-photo-filter/ id404497747?mt=8>), BeFunky Photo Editor by BeFunky (<http://itunes.apple. com/us/app/befunky-photo-editor-pro/id440241836?mt=8>), FX Photo Studio by MacPhun LLCHD by (<http://itunes.apple.com/us/app/fx-photo-studio-hd/ id369684558?mt=8>), WoodCamera by Bright Mango (<http://itunes.apple.com/ us/app/wood-camera/id495353236?mt=8>), and Grungetastic by JixiPix Software (<http://itunes.apple.com/us/app/grungetastic/id418140198?mt=8>).

There are also picture Graphics Interchange Format (GIFs), which are short video loops—no more than one of two seconds long; they include Tumbir (<https:// www.tumblr.com/>), Cinemagram by Factyle (<http://itunes.apple.com/us/app/ cinemagram/id487225881?mt=8>), Flixel by Flixel Photos Inc. (<http://itunes. apple.com/us/app/flixel/id496885363?mt=8>), Gif Shop by Something Savage (<http://itunes.apple.com/us/app/gif-shop/id410174605?mt=8>), and Gifboom by TapMojo LLC (<http://itunes.apple.com/us/app/gifboom-animated-gif-camera/ id457502693?mt=8>) and are an easy way to create moving picture effects.

Finally, there are movie apps that can be used to create videos with editing, shooting, and social sharing features, including iMovie by Apple (http://itunes. apple.com/us/app/imovie/id377298193?mt=8; Figure 2.12): There are hundreds of mobile apps available on the market, from games through to photo editing and sharing applications, many of which are available through the Apple App Store. It is a well-known fact that the 2-megapixel cameras that Apple mobiles support are not the best cameras around, and even more disappointingly is that they cannot capture video recordings. However, there are some great third-party apps that can

Figure 2.12   The iMovie app by Apple®.

enhance the mobile photography experience by enabling users to edit and share their photos easily with family and friends.

There are also some ingenious apps that enable users to capture video and share them with their friends. For the developer, these video apps are a great vehicle for broadcasting ads. Some of the best include isurp8 by MEA Mobile (<http://itunes. apple.com/us/app/isupr8/id413566476?mt=8>), 8mm Vintage Camera by Nexvio Inc. (<http://itunes.apple.com/us/app/8mm-vintage-camera/id406541444?mt=8>), SilentFilmDirector by MacPhun LLC (<http://itunes.apple.com/us/app/vintagio/ id335148458?mt=8>), VintageVideoMaker by MacPhun LLC, Vimeo by Vimeo, LLC (<http://itunes.apple.com/us/app/vimeo/id425194759?mt=8>), mogo video by Hansel Apps (<http://itunes.apple.com/us/app/mogo-video/id419439294?mt=8>), Precorder by Airship Software (<http://itunes.apple.com/us/app/precorder-video-camera-for/id412558814?mt=8>), FilmicPro by Cinegenix, LLC (<http://itunes. apple.com/us/app/filmic-pro/id436577167?mt=8>), VideoCamera+ by Totus Pty Ltd. (<http://itunes.apple.com/us/app/video-camera+/id441433868?mt=8>), and ReelDirector by Nexvio Inc. (<http://itunes.apple.com/us/app/reeld-irector/id334366844?mt=8>). There are also some social video apps such as Socialcam by Jusint.tv (<http://itunes.apple.com/us/app/socialcam-video-camera/ id421228047?mt=8>), Klip by Klip Inc. (<http://itunes.apple.com/us/app/klip-video-sharing/id445539290?mt=8>), Cinecam byyuzamobile.com (<http://itunes. apple.com/us/app/cinecam/id445584503?mt=8>), and Viddy by Viddy, Inc. (http:// itunes.apple.com/us/app/viddy/id426294709?mt=8; see Figure 2.13).

Figure 2.13   Social video app reportedly worth $375 million.

2.13 Gov Apps

Recently, President Obama issued a directive that aims to bring government agencies into the twenty-first century by making such things as apps a top priority going forward. Obama argued that the American people have been forced to navigate "a labyrinth of information across different Government programs" (GAO) in order to find the services they need, bemoaning the government's slow adoption of technologies that consumers have come to expect in the age of mobiles.

"At a time when Americans increasingly pay bills and buy tickets on mobiles, Government services often are not optimized for mobiles, assuming the services are even available online," Obama wrote. The President referenced Executive Order 13571 (Streamlining Service Delivery and Improving Customer Service; <http://www.whitehouse.gov/the-press-office/2011/04/27/executive-order-streamlining-service-delivery-and-improving-customer-ser>), issued on April 27, 2011, which required executive departments and agencies to, among other things, identify ways to use innovative technologies to streamline their delivery of services to lower costs, decrease service delivery times, and improve the customer experience.

That order led to the creation of a continuing effort by the government that will enable more efficient and coordinated digital service delivery by requiring agencies to establish "specific, measurable goals for delivering better digital services." Agencies will be required to make information readable across mobiles and open to app developers for the creation of more user-friendly services and information.

Ultimately, this strategy will ensure that federal agencies use mobiles to serve the public as effectively as possible. Americans continue to rely on their mobiles for anytime and anywhere access, including the Internet. President Obama, a known BlackBerry and new iPhone user, sounded a little like an advocate for the wireless industry, noting in a statement that: "Americans deserve a government that works for them anytime, anywhere, and on any device (GAO)."

2.14 A Stealth App

A novel approach to app ads is being provided by MobilePosse (<http://mobileposse.com/>). They proactively deliver information, entertainment, and marketing content to the *idle* screen of mobiles. Graphical and interactive messages are delivered only when the mobile is not already in use, thus ensuring that calls, texts, and browsing are not interrupted. Content includes weather forecasts, sport scores, and trivia, as well as exclusive coupons and other discounts in a nonintrusive manner—all amenable to strategic mobile analytics and marketing.

Mobile Posse, Inc., is an advertising platform for the active home screen. Using proprietary patent-pending technology, Mobile Posse enables advertisers, content providers, and wireless carriers to proactively reach consumers through the prime real estate on the mobile phone. The company provides mobile content

Figure 2.14 Mobile Posse.

and advertising solutions. Their app solutions enable consumers to receive discounts, offers, and content on mobiles, as well as allows advertisers to deliver targeted and media advertising to consumers. Mobile Posse also enables mobile operators to engage opted-in subscribers through targeted messages and allows content providers to deliver targeted content to consumers that opt in to receive offers in a nonintrusive manner.

Mobile Posse (http://mobileposse.com/) is a free service that keeps users in-the-know, wherever they go, with weather forecasts, local gas prices, sports scores, and more—along with great savings from national and local retailers. The app delivers informative and fun content to mobiles when in use or not. Mobile Posse allows for the setting of profiles and preferences to specify the type of content and offers users want to see (Figure 2.14).

2.15 Apps Overtake PCs

The notion of the post-PC era may have legs to stand on, and Facebook could be playing a key role in the shift. However, increasing apps usage comes at the expense of desktop and laptop devices; if the social network does not respond quickly, others will eat their lunch. Mobile apps continue to rise in popularity, with people now spending 94 minutes per day, on average, using apps, according to analytics firm Flurry (Flurry.com). Meanwhile, Web consumption, on both desktops and laptops, has started to wane. The mobile analytic firm Flurry tracked anonymous sessions across more than 140,000 apps and compared that data against Web data from comScore and Alexa. Since the Flurry's initial report, mobile app usage has increased by 13 minutes a day, while Web desktop consumption dipped from 74 minutes to 72 minutes per day in 2011.

What are mobile app users doing? Principally games (49 percent) and social networking (30 percent), which combine to account for almost 80 percent of all time spent in apps, according to Flurry's analysis. The growing chasm between app usage and Web consumption will only be exacerbated by mobile purchases in 2012 and the coming years. Flurry estimates that the cumulative number of Apple and Android mobiles activated will surge past 1 billion. According to IDC, over 800 million PCs were sold between 1981 and 2000, making the rate of Apple and Android mobiles adoption more than four times faster than that of personal computers.

Flurry anecdotally attributes the measurable shift to Facebook users who are increasingly accessing the social network via apps. The firm does not track Facebook usage, but points to Nielsen data on Facebook mobile usage and the success of Facebook Messenger to support its hypothesis. As Apple and Google continue to battle for consumers through the operating systems and mobiles, Facebook is demonstrating that it can leverage its hold over consumers at the software level, through the power of their social network, across multiple platforms. So maybe it is not evidence of a post-PC era per se, but instead a Facebook era that has consumers clamoring for access to their News Feeds wherever they are. Here is another interesting take-away that may put to the rest the native versus HTML5 app debate: The decrease in Web consumption, which includes mobile Web usage, could be evidence of consumers' growing preference for rich, native application experiences.

2.16 Health Apps

The mobile health app market will grow to $392 million over the next five years—a 70 percent increase—according to new data from research firm Frost & Sullivan (<http://www.healthcareIT.frost.com>). And it might be quite a bit more than that, as the market has consistently outpaced forecast growth and revenue over the past two years, according to the firm's report. Health apps will continue on a steep growth curve as increasingly sophisticated mobile technologies and relationship-management tools disrupt the market. Despite the hype, mobile apps are the single-largest digital channel since the 1990s and the Web. Frost researchers predict that not only will new users buy health apps, but also that existing customers will continue to buy and use more health apps.

Calling 2011 the "tip of the iceberg," researchers say that low barriers to market entry will continue to lure new vendors, although they note that U.S. Food and Drug Administration oversight and security concerns mean it will not be a hassle-free endeavor for new start-ups. Consumer awareness is mixed, with privacy and security concerns ever-present in the health market. Also, while still overall a good thing, increases in FDA (Food and Drug Administration) regulation and oversight may dampen innovation. A related Frost report (Frost.com) also forecasts growth in the remote patient monitoring market, with a strong mobile component. Frost projects revenues will more than double to $295 million in 2016, up from $127 million in 2010. Interestingly, the report predicts that remote patient monitoring technology will move in a more consumer-facing direction, driving much of that growth. Even with a double-digit growth rate for the past decade, Frost researchers say the market has not reached its "billion-dollar potential" because of scalability problems and limited business models.

2.17 Building a Popular App

In today's noisy app industry, there are very few companies that can simply build an app, generate downloads, and see continual adoption and user engagement. And, with the burgeoning U.S. mobile industry, companies had better determine how to build the most engaging app imaginable. App developers and marketers can no longer assume that they can merely hop on the mobile bandwagon and reap immediate rewards. In fact, companies must think beyond the initial download in order to drive continual user interest.

Companies can now build engaging apps accessible from anywhere in the world, more than doubling their consumers by simply providing terrific content for every mobile platform. In order to ensure that mobile users come back for more, there are five basic steps to follow during app development and post download.

1. *Identify user demand, rather than pushing corporate marketing; the branding benefits will come if the app developer acknowledges this fact.* In developing their apps, companies repeatedly spend tens of thousands of dollars and months of personnel time and resources, only to launch the apps and see usage rates drastically drop. The reason is often unanimous: Rather than building an app to serve a market need, brands, developers, and organizations should build apps that serve themselves, hoping that users will care enough to download. Unfortunately, unless there is a clear value to users, apps will be ignored. Identify what makes your company, products, and insights truly unique. Consider how this information can be delivered quickly and easily, and what incentives can be added to reward ongoing use of a mobile app.

2. *Brands and companies should assign responsibility and empower app managers.* Because the mobile industry is rapidly growing, there is a need for brands to put their best people on mobile. For instance, all major digital communication technologies are headed to mobile—telecoms, computers, and the Internet. In fact, by 2015, mobile advertising is slated to be a $24 billion

industry. In addition, all major media channels are embracing mobile—whether it is music, gaming, news, television, or advertising. While a strong PR (public relations) plan can support this effort, brands and companies must delegate full responsibility to an individual, team, or department in order to ensure care for the mobile community. Furthermore, mobile updates and new content must continually refresh the app and the user experience. Marketers and app developers must implement an ongoing plan to not only continue the number of downloads, but also sustain and increase the amount of time consumers spend using the app.

3. *App developers need to identify their target audience and implement the features they are most likely to use; this will ensure that the data mining of mobiles will be beneficial to their brand and organizations.* The mobile coupon industry, for example, is slated to be worth $46 billion by 2016 according to Flurry. com. Developers need to give their app users a chance to "opt in" to updates and implement ways to incentivize and to reward certain mobile behaviors. Tap location-based features that display when and where your users are most likely to engage with brands and enterprises; developers should integrate social media widgets that allow them to learn about, engage with, and reward users directly on their mobiles.

4. *Developers should build an app and a plan that focuses on relevance and, above all, consumer engagement.* The three guidelines for a winning app are relevance, great functionality, and a slick user interface. As mobile marketers are learning, engagement rates trump all other analytics and are the most relevant at determining which marketing tactics are effective, which stimulate user engagement, and which drive sales or calls-to-action. Providing relevant content requires analytics, diligence, and continual updates based on continuous metrics.

5. *App developers need to study their analytics and adjust accordingly.* It is important to monitor which features generate the greatest use and understand what content users share, what offers and ads users ignore, and which live radio or video streams receive the most attention. Minding the analytics about user engagement with your app will help determine which features need improvement, investment, or deletion.

As the app industry continues to flourish, marketers must take chances and test the adoption and use of new features.

Finally, mobile app development can be used for branding an organization or business. As mobiles are gaining popularity, the demand for various apps is also increasing.

Mobiles have opened up new areas to be visited by firms; and even with the fast growth of the business branding market, it can be said that the true prospective of mobiles on the market is not comprehended as yet. The surge of the gaming market has elite the promotion professionals to use social gaming to set up company popularity. Marketing information is often sent along with the programs and games in a simple way so that clients get the concept, but are not irritated by it.

When enterprise manufacturers recognized that long-term company technique can be established with apps that are relatively easy to set up by devoted designers and mobile clients are used to buying from mobile apps shops immediately. Rather than basically avoiding mobile-enhanced websites, firms now want to take one big step ahead and create mobile apps. The practical knowledge provided by mobile programs is completely different from a mobile website. Apps are highly entertaining and can be used to actively interact with consumers to discover their needs and desires. They can deliver a continuous stream of consumer information for brands and enterprises.

2.18 An App Ad Reporter

Another innovative app platform is Medialets (<http://www.medialets.com/>) (Figure 2.15); it combines rich media advertising with behavioral analytics and supports Apple and Android mobiles. Medialets provides a platform for brands and agencies to create ads for all mobiles. Their Enrich™ program enables mobile ad networks, as well as ad mediators and ad servers, to sell and support their cross-platform rich media units. Advertisers can run a single set of ads without changes across a broad range of mobiles and access a unified set of reports from which to measure marketing campaigns. Medialets gives developers and marketers insight into the performance of their app usage and campaign performance.

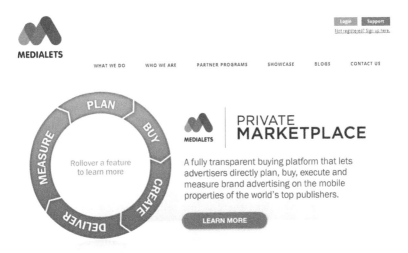

Figure 2.15   The Medialet platform for mobile metrics.

Their app metric reports are available through their Medialytics, an online reporting tool that provides real-time data, charts, and graphs, as well as the ability to set custom reporting filters and email alerts. Their app reports provide number of visits, unique visitors, session length, and other custom metrics. Medialets provides a 360-degree look at mobile marketing campaign performance, including impressions, clicks, engagement and engagement rate, and custom campaign metrics. Third-party ad verification and add-on brand studies are also available. Because Medialets is cross-platform, it is possible to compare metrics across Apple, Android, and other major and popular mobiles. Medialets tracks app activity and campaign performance even when the device is offline.

2.19 Biz Apps

Apps are important to a brand or a company because they make even small businesses look big by enabling them to keep their new and existing customers engaged with their products and services. Business apps offer new innovative ways to reach consumers. For example, Domino's Pizza® (<http://www.dominos.com/>) has the PizzaTracker app that allows customers to track their orders and deliveries. For this app, the company won the prestigious Webby Award (<http://www.webbyawards. com/>). These kinds of success stories are inspirational for small businesses. But, before getting and constructing an app for a brand or a business, there are several factors to consider.

Apps are software that requires extensive man-hours to develop. The need of each brand and business is different. Developers need to consider their business issues and customer requirements, budget, etc., when choosing to create a mobile app. However, an inexpensive and simple apparatus such as Magmito (an online app creation tool; <http://www.magmito.com/>) (Figure 2.16) can be used by a small business to create an app and to distribute it to any mobile operating system on the market to create a click-to-call app and one that offers special deals.

Small businesses need to consider who their customers are and how and when they use their mobiles to reach them. They need to take care of and provide fresh content for their apps. If they are unable to develop new content for their apps, all the efforts will go waste and they will lose their customers. Consideration must be given to what mobiles their customers use. Different mobile platforms need different apps that should be able to solve the problem of consumer, but also be easily accessible to them.

For business app developers, there is one reliable tactic they can employ that will not only boost their notoriety, but also improve their revenue: give it away. Despite the concept sounding counter-intuitive—by lowering the price of an app, they can actually make more money; the numbers have been crunched by people in-the-know. In a study across the three main app stores—Apple App Store for iPhone, App Store for iPad, and Android Marketplace—Distimo (Magmito.com)

Figure 2.16   Business app creation tool.

found that despite selling an app at a reduced price, the surge in downloads makes up for the loss in price, enough to actually turn a profit on the less-expensive app.

Giving away an app not only encourages more downloads from users, but it will also get them featured in these app stores. Distimo provides valuable insight into the app store markets for developers, carriers, and device manufacturers. While free apps generated an overall increase in revenue, only two-thirds of Apple apps gained rank in the first 3 days after being featured. However, having an app featured does not automatically guarantee a boost, but it still is likely to help.

The concept behind a great app may be groundbreaking, but if its underlying mechanics are faulty, no amount of marketing will improve its reputation. Putting an app full of bugs on the market will not only land the developer with discontented customers in the short-term, but will also affect its long-term reputation, even after the glitches are fixed. The only answer to this is effective and rigorous testing. The app will never achieve widespread success if the brands or enterprises are the only ones who know how fantastic it is. An effective marketing campaign will maximize the visibility of an app's products and services—helping the app stand out in the overcrowded marketplace is something the brand's developer must be prepared to devote time and funds to.

As a developer, it is disastrous to rely on the customer to come to the app unprompted, and complacency about marketing it properly can result in it being overlooked, no matter how good it is. Once an app is perfected, do not let the app sit gathering dust in a dark corner of an app store. A snappy free product description is an opportunity to capture the attention of potential customers; make it creative, exciting, and intriguing. Likewise, setting up a website for the app provides the opportunity to give app enthusiasts more detailed information about it, including screenshots, news on future developments, and even troubleshooting advice.

This applies to both the app's audience and its competitors. It is easy for enthusiastic developers to become preoccupied by an idea without sparing a thought for potential customers. Having a targeted audience in mind—and tailoring marketing efforts to their interests—is vital to successfully growing the reputation of the app. Similarly, being aware of competitors allows developers to avoid the embarrassment of their apps not being "one-of-a-kind" as initially envisioned.

There are numerous methods for marketing an app, and some are more effective than others. Advertising, for example, can take the product to its intended audience, but for new developers who lack the budget, it can severely deplete available funds. If the developer is still keen on using advertising, consider a pay-per-success campaign where they are only required to part with money if the advert leads to a direct download. This is often cheaper than paying for upfront advertising and means the brand or enterprise can track the direct downloads.

Word-of-mouth is perhaps the best way of creating a trusting relationship with app users; and although persuading journalists and bloggers to write about an app can involve perseverance, it is well worth the effort. Similarly, user-generated feedback can drastically affect download rates. The message with regard to effective app marketing is that of a sustained, measured approach as the best way to guarantee continued success. Taking the time to ensure the quality of the product, considering the market, and designing a marketing strategy is vital. Developers should not spend their entire marketing budget in the first few weeks. While expensive advertising campaigns can generate an early spate of downloads, they do not ensure the app's sustained success. A combined approach incorporating word-of-mouth, viral marketing, and advertising is the best was to keep downloads high and the app's profits secure.

Mobile apps are a large part of our everyday world. The most popular mobile apps for the coming years will have unique features that cater to the mobile environment rather than merely act as an extension of their online peers. Mobile applications will be a highly competitive marketplace this year, and those device vendors that proactively integrate innovative apps at the platform layer will have the competitive edge. Here are a few of the cutting edge mobile application trends for the near future:

1. Location-based services deliver features and functionality in tune with their end user's context, taking into account the user's location, personal preferences, gender, age, etc., and thus offering a more intelligent user experience.

An example of location-based services would be Google Maps and Foursquare (<https://foursquare.com/>).

2. Mobile social networking is the fastest-growing consumer mobile app category; it allows end users to share videos, photos, games, emails, and instant messaging via mobile devices.

3. Mobile payments will not become mainstream soon; there will be an increase in the amount of apps available. Developers still have to work through usability, ease of implementation, and security for all end users to jump on board.

4. Mobiles with larger screens offer the ideal platform for video consumption. Within the past few years, the use of video has grown from a few businesses occasionally including it on their websites, to an essential, competitive differentiator that can streamline operations and quickly promote a brand via apps.

5. Mobile search is usually related to a product, services, or places search. Currently there will be more apps available to render purchasing products, making reservations, buying a ticket, etc., easier for the end user.

2.20 A Video App

Break Media (<http://www.breakmedia.com/>), a leading creator and publisher of digital video content, found that digital video ads spending is up and is expected to continue to increase for the next several years. In addition to organic growth of budgets, the increase is projected to come from new platforms such as mobile apps—easily the fastest-growing format; connected mobiles and their apps are the dominant new emerging ad format.

The company's properties include the largest humor site online –(Break.com), as well as Made Man, Game Front, Holy Taco, Screen Junkies, Cage Potato, All Left Turns, Chickipedia, and Tu Vez. The Break Media Creative Lab is an in-house production studio that creates original videos ranging from award-winning branded entertainment to celebrity-driven Web shorts to viral one-offs. The Break Media Network represents hundreds of publishers as one of the largest video advertising networks online reaching more than 140 million visitors each month.

Mobile app video ad spending is increasing; in the coming year, 68 percent of advertisers will increase the share of online display advertising via apps. Video ad network use is skyrocketing, with more than 90 percent of all advertisers planning to use them in the coming year, thus increasing the share of spend devoted to them from an average 20 percent to 41 percent of total video dollars.

The cost-per-view (CPV) model offered by more and more publishers and networks, and the pervasive use of video ad networks (VANs), has driven variety in the pricing models available, and the CPV model has increased two-fold in the past year. The CPV model has clearly caught traction in a short amount of time, and new formats (including mobile video and apps advertising on mobiles) will be compelling ways for brands to connect with consumers in the coming years.

2.21 The Twitter Apps

Twitter is not only a relatively new social media real-time network feed, it is also a new ad platform for targeting mobiles. Twitter allows for marketers to promote and calibrate the effectiveness of their advertising tweets—including important feedback on how many people clicked on their links. Twitter provides vital metrics such as reuse hashtags, retweet of a post, or marks as a favorite, or the number of mobiles that choose to follow a thread. Retweets are cascading tweets to other mobile followers on Twitter; they allow for an immediate method of measuring the marketing of a brand—or the effectiveness of an ad campaign, which for app developers is a whole new market.

For mobile marketers and brands, there are thousands of apps for leveraging Twitter; just Google "Twitter apps" to see the most current ones. For example, one Twitter service is HubSpot (<http://www.hubspot.com/>), which can be used to provide quality leads to mobiles. The service analyzes when followers tweet and then recommends the best times for a brand or marketer to tweet; it is fairly simple and free.

Twitter has also pushed out a new layout that introduces three new buttons to the mobile marketer: they are Home, @Connect, and #Discover. The Home button takes the user to their main Twitter feed. The @Connect button shows the user all the interactions he or she has had with other users, such as favorites, conversations, etc., and displays suggested accounts. The #Discover button features trending news stories and new hashtags.

A new Twitter app, known as TweetDeck (<http://www.tweetdeck.com/>) is an indispensable tool for tracking real-time conversations about any given topic. For example, direct messages, saved searches, personal lists, and profile information can be captured. Built with HTML5, the Web version of TweetDeck syncs user accounts, columns, layouts, and settings whenever and wherever users sign in. TweetDeck reflects the overall design of Twitter with Profile and Tweet box pop-ups. TweetDeck provides a simple way to filter content. Users can easily set up columns that track search results, trending topics, specific users, their own @ mentions, favorites, and more. Users can also open Tweets directly within a column to view media, conversations, favorites or retweets; they can also choose to post an update at a specific time in the future using TweetDeck's scheduling feature.

Another popular Twitter app is UberTwitter, which can be used by mobile marketers to insert ads into users' Twitter streams. UberTwitter is now UberSocial (<http://ubersocial.com/>), to reflect its broadening reach as a key social communications tool. UberSocial can be used to send and read tweets, and to post rich media to Facebook. Another popular social marketing media app is Seesmic (<https://seesmic.com/>), which can also be leveraged by mobile analysts and marketers for broadcasting via Twitter. Seesmic focuses on building apps to help users build and manage their brands online; it specializes in social media monitoring, updating, and engaging in real-time.

There is also a new company, Klout (<http://klout.com/home>), that can provide Twitter metrics and mobile analytics for developers and marketers, enabling them to identify the Twitter users who are most influential in online discussions about their brand. Klout measures the influences of mobiles across the social Web and allows marketers to track the impact of opinions, links, and recommendations of restaurants, movies, bands, games, brands, etc. Klout allows mobile marketers to find influencers based on a topic or a hashtag. When a mobile sees a pound sign (#) followed by the name of a product, service, or a brand, it is seeing what is known as a Twitter Hashtag or Tag. Hashtags offer an easy way to perform a keyword search on a topic or to define a posting under that topic on Twitter.

2.22 App Usage Patterns

Nuance Communications (http://shop.nuance.com/>) recently released a survey that shed light on consumer usage of mobile apps and their impact on companies' customer service strategies. Nuance is a provider of voice and language solutions for businesses and consumers. Its technologies, applications, and services make the user experience more compelling by transforming the way people interact with mobiles and systems.

The survey found that companies hoping to engage with their customers through mobile apps must not only consider how they attract first-time users, but they must also consider retention strategies and how they keep consumers returning regularly to this critical channel. Mobile apps are an increasingly important channel for customer service. When you consider the forecast from Nuance.com that nearly 2 billion people will have mobiles by 2017, coupled with the fact that consumers spend nearly 10 percent more time using mobile apps than they spend on the Internet, it is clear why there is significant opportunity—if not a strategic imperative—for companies to leverage the mobile channel for customer service.

The survey found that the benefits of companies providing a mobile app for customer service uses are twofold: (1) company perception and (2) customer satisfaction. The majority of mobile users surveyed (72 percent) said that they have a more positive view of a company if they have a mobile app. And they will not keep it to themselves. Eighty-one percent will tell others about a positive app experience. Weighing in on how mobile apps would increase their levels of customer satisfaction, 35 percent said that effortless transition to a live agent from a mobile app is the feature most likely to drive mobile app usage, while 48 percent want more: offer more functionality (21 percent) or find other ways to "better meet my needs."

According to the survey, 89 percent of mobile users download at least one new mobile app per month, signaling that consumers are willing to download and try new apps. Seventy percent of consumers surveyed said they have more than ten apps on their phones, while 29 percent have more than thirty apps on their phone, and another 12 percent have more than fifty. While the download number continues in an upward pattern, a recent study from Flurry (Flurry.com) found that mobile apps are quickly abandoned: after just 3 months, only 24 percent of consumers continue to use an app they have downloaded, and after 12 months' time, the majority of apps (96 percent) are no longer used. Companies hoping to engage with their customers through mobile apps must not only consider how they attract first-time users, but they must also consider retention strategies and how to keep consumers returning regularly to this critical channel.

While consumers continue to initiate their communications with customer service primarily through traditional channels, such as the phone and Web, mobile apps are becoming an increasingly important customer engagement channel. The uptake varies vertically; more than half of mobile users have downloaded their bank and their carrier's app, with 45 percent of users stating, "It's simply more convenient." But just because they have downloaded the app does not mean they use it. The survey (Flurry.com) found that of carrier apps, 60 percent of mobiles users have downloaded the app; of those surveyed, only 25 percent of them use it. On banking apps, 55 percent of mobile users have downloaded the app; of those surveyed, only 27 percent of them use it. Finally, regarding brokerage apps, 27 percent of mobile users have downloaded the app; of those surveyed, 18 percent of them use it.

Another survey by research firm Ovum (Ovum.com) reported that consumer expectations for seamless self-service are shifting. What once was considered a "nice-to-have" is now a "must-have." Consumers have come to expect the ability to easily move from their mobile device to other channels.

Companies that can successfully deliver this multi-channel experience will be able to differentiate themselves from the competition, and more importantly better serve their customers. While mobile apps are playing an increasingly important role in customer service, the survey results underscore the opportunity for companies to differentiate by delivering an integrated, highly personalized, cross-channel customer experience.

Today's consumers look for convenience, and they are shifting their spending to brands and companies that keep them better informed and are easier to do business with. Providing a powerful mobile experience is central to building customer loyalty. Another consumer survey, conducted by Vocalabs (Vocalabs.com), which included a random sample of 991 mobile users in the United States and Canada, to get a better understanding of their mobile app usage and appetite for customer self-service through their mobiles, found that apps play an important role in self-customer service.

2.23 The Whispering Apps

Social media marketing via friends is about word-of-mouth (WOM) that can be encouraged and facilitated via Twitter and social networks. Mobile marketers and their clients can communicate with mobiles by making it easy for them to tell their

Figure 2.17   The bango® platform.

friends. Twitter targeting can make certain that influential individuals know about the good qualities of a brand, a product, or a service, and make it easy for them to share it with other mobile devices in a relevant manner via WOM.

Flurry (<http://www.flurry.com/>) specializes in mobile apps analytics; thousands of enterprises and brands use their technology in more than 100,000 applications across Apple, Android, Blackberry, Windows Phone, and J2ME. Their Flurry Analytics service helps developers deepen consumer engagement and improve the monetization of their apps. The service is free, cross-platform, easy to integrate, and able to handle data loads of any size. Then there is bango® (<http://bango.com/>), which does a little bit of both; their analytics can report on mobile site navigation and traffic as well as app use (Figure 2.17). Their platform operates on a white label basis and can be used by content providers and app developers to sell directly to consumers.

Here is almost everything you need to know before you get started on your own app. For example, making an app can cost anywhere from less than $100 at the very minimum, to up to $10,000. The minimum price is for a super-simple program with none of that fancy enterprise or any social networking features. Unless the mobile marketer has some basic design skills, he or she will need to enlist the help of both an HTML5 programmer and an experienced app designer, which can raise the cost. One more unavoidable cost is Apple and its charge of $100 per year to hold onto a developer's account, which a business or brand needs in order to publish their app to the world.

The next consideration is the app price: consider at or near $1.99. This is the premium price, but it is also immensely satisfying to get more than a buck per download after Apple takes away its 30 percent. And, as with most things, it is a lot easier to lower the price later than it is to raise it. The next consideration is simply giving away the app to potential or existing customers. If the primary business model involves in-app purchases, ads, or the like, developers probably will want to give away their apps for free. A brand or business should take a quick glance at Apple's top grossing charts that show the popularity of free apps.

Getting an unreleased app tested is not the easiest thing in the world; however, app developers can use a program such as TestFlight (<https://testflightapp.com/>), which makes it very easy to test and build updates to register one's apps. Getting featured on iTunes is obviously awesome; when Apple includes an app on its featured lists, developers will enjoy a predictable flow of downloads almost identical in volume every single day. Especially fascinating, the "New & Noteworthy" list provides almost exactly twice as many daily downloads as the "What's Hot" list. This is because, when you tap on the "Featured" tab on the "App Store" app, "New & Noteworthy" pops up by default.

2.24 Hyper Targeting Apps

The success of mobile modeling involves strategic planning and measured improvement of predictive evolving clusters. The key objective for executing and leveraging mobile analytics is to plan and design a framework from which consumers' behaviors can be captured and predicted at mobile sites and app usage. Similarly, this mobile modeling strategy should be to create a continuous and systemic method of quantifying mobile behaviors and to continuously measure everything. The modeling of mobile apps has three different requirements:

1. What is the purpose of the app itself and its structure? For example, in the case of Groupon, it is offering discount deals based on the location of the mobile.
2. What is the business logic of the app and its components? Is the app based on location, interest, proximity, age, gender, time-of-day, or a combination of some or all of these factors, as is the case with Pandora?
3. Finally, the graphical user interface of the mobile app is vitally important: it should be simple, functional, and slick.

One of the most popular operating systems for apps is Android and its platform from the Open Handset Alliance (<http://www.openhandsetalliance.com/>), whose thirty-four members include Google, HTC, Motorola, Qualcomm, and T-Mobile. It is supported by all of these major software, hardware, and telecom companies. It uses the Linux kernel as a hardware abstraction layer (HAL). App development and programming is mostly done in Java. The Android-specific Java software development kit (SDK) is required for development, although any Java

IDE (Integrated Development Environment) may be used. Performance-critical code can be written in C, C++, or other native code languages using this Android native development kit (NDK).

When it comes to mobile app advertising, everyone agrees that someday the market will be huge—if only the participants could figure out a way to bring reliable analytics to the incredibly complex and highly random act of using mobiles. For the mobile sports market, a company called CrowdOptic (<http://www.crowd-optic.com/>) is trying to crack the code with a technology base that can offer real-time mobile analytics about what people at a game are watching to advertisers, teams, and other interested parties, while also providing a real-time communications stream back to mobiles that could significantly enhance the sporting event they are attending.

This start-up has already demonstrated its ability to use its unique triangulation algorithm and augmented-reality app to give fans at a sports event real-time information about the player they have just snapped a picture of. On the back end, CrowdOptic is able to give event organizers detailed information on exactly what the most fans were looking at through their mobiles, a practice the company calls "hyper targeting," which theoretically could provide incredibly granular sets of data about what exactly is catching people's attention at a sporting event. It is all wrapped under a banner that the company calls "Focus-Based Services," in an attempt to move the discussion beyond location-based services and to a place where users can determine what other people are looking at, and not just where they are.

The CrowdOptic technology can detect, in real-time, where a crowd of mobiles are shifting their attention, and are looking into the possibilities for the platform to support other applications such as stadium security, in-seat advertising, and ticketing. The crown jewel of the start-up is their algorithm of triangulation. With a small app installed on a mobile, CrowdOptic takes information from the device's GPS service and its camera and feeds it into their system that can then provide analytics to show what other fans are pointing their devices at.

2.25 Crashing Apps

In a new study (Crittercism.com) conducted by the mobile application monitoring company Crittercism, it is noted that iOS applications crash more often than their Android counterparts. The results speak for themselves: more applications crash under iOS 5.0.1, while Android apps appear to be more stable. However, we should note here that the faults regarding iOS 5.0.1 most likely do not lie with Apple itself. Because this version of the mobile operating system is relatively new, it is more likely that iOS developers have not made their own third-party apps fully compatible with the iOS, thus resulting in the crashing. On the other hand, iOS 4.3.3, which has indeed been available for some time, also caused a sizable 10.66 percent of mobile apps to crash.

Android, as you can see, is quite stable; its most recent Android 4.0.1 accounts for just 1.04 percent of mobile app crashes. Compare that with iOS 5.0.1 at 28.64 percent!

If you are wondering why iOS applications crash so much, there are a number of potential reasons. This can be due to hardware issues, such as the use of location or GPS services or cameras; it could be due to the Internet connection, that is, how a mobile connects to 3G or Wi-Fi, or that the device is not connected to the Internet at a certain moment, or that something happens during the switch between 3G and Wi-Fi. There could also be issues with language support on certain mobiles. There can also be memory problems if an app uses too much memory. Furthermore, Apple's iAd system is a potential problem; it apparently gives developers problems if they do not adhere to certain standards. But that is not all; there is also the common problem of developers not updating their apps.[*]

2.26 Mapping Apps

Just as MapQuest kicked off a rush to provide street-by-street navigation for Internet users a decade ago, the race is now on to create maps via apps for stores, airports, malls, arenas, and other indoor venues. Start-ups such as aisle411 Shopping Companion by aisle411 (<http://itunes.apple.com/us/app/aisle411-shopping-companion/id394218369?mt=8>), Micello (<http://www.micello.com/>), and Meridian (<http://www.meridianapps.com/>) are all developing mapping apps for mobiles. Users can request that their favorite store be "mapped" by these start-ups, which can help them navigate inside a large retail store. Google (Guidebook.com) is already providing maps for United States airports, as well as hundreds of Home Depot and Ikea stores via their Android operating system.

These types of mapping apps can be used to learn the paths consumers take and what items they are looking for, as well as what purchases they make. It is a new form of mobile "market basket" analysis. Market basket analysis is a modeling technique based on the theory that if consumers buy a certain group of items, say cheese, they are more or less likely to buy another group of items, like wine. Clustering and association algorithms make mobile market basket analysis possible, and the technique and software will be covered in detail in Chapter 5. These mapping apps monetize these mobile behaviors inside retail venues; firms such as Guidebook (<http://guidebook.com/>) have created maps for museums, industry conferences, and other retailing indoor spaces (Figure 2.18).

---

[*] According to Tomio Geron of *Forbes*, *Forbes* online 2/2/12.

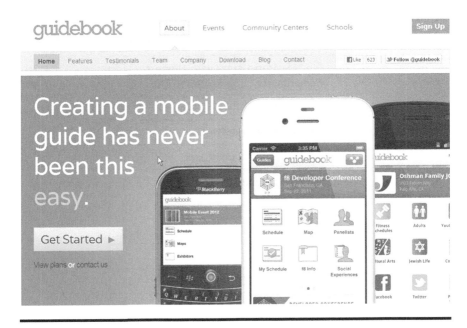

Figure 2.18   A platform for mapping large indoor events.

2.27 The American App Economy

TechNet, the bipartisan policy and political network of technology CEOs that promotes the growth of the innovation economy, recently released a new study (Technet.com) showing that there are now roughly 466,000 jobs in the "App Economy" in the United States, up from zero in 2007. The study, sponsored by TechNet, also found that App Economy jobs are spreading throughout the nation. The top metro area for App Economy jobs is New York City and its surrounding suburban counties, although San Francisco and San Jose together substantially exceed New York. And while California tops the list of App Economy states with nearly one in four jobs, states such as Georgia, Florida, and Illinois get their share as well. In fact, more than two-thirds of App Economy employment is outside California and New York. The results also suggest that the App Economy is growing quickly, and that the location and number of app-related jobs are likely to shift significantly in the years ahead. App brands, marketers, and analysts should be aware of where these developers are because employing those who are local is an important consideration.

America's App Economy had zero jobs just a few years ago before the iPhone was introduced; however, its introduction demonstrates that such mobiles can quickly create economic value and jobs through cutting-edge innovation. Today, the App Economy is creating jobs in every part of America, now employing hundreds of thousands of U.S. workers and even more in the years to come. The App

Economy, along with the broad communications sector, has been a leading source of hiring strength in an otherwise sluggish labor market.

Top U.S. Metro Areas with Highest Percentage of App Economy Jobs	
New York–Northern NJ–Long Island	9.2%
San Francisco–Oakland–Fremont	8.5%
San Jose–Sunnyvale–Santa Clara	6.3%
Seattle–Tacoma–Bellevue	5.7%
Los Angeles–Long Beach–Santa Ana	5.1%
Washington–Arlington–Alexandria	4.8%
Chicago–Naperville–Joliet	3.5%
Boston–Cambridge–Quincy	3.5%
Atlanta–Sandy Springs–Marietta	3.3%
Dallas–Fort Worth–Arlington	2.6%
Top Ten States for App Economy Jobs (Percentage)	
California	23.8%
New York	6.9%
Washington	6.4%
Texas	5.4%
New Jersey	4.2%
Illinois	4.0%
Massachusetts	3.9%
Georgia	3.7%
Virginia	3.5%
Florida	3.1%

The research shows that when it comes to employment impacts, each app represents jobs for mobile programmers, user interface designers, marketers, managers, support staff, and most importantly for analysts and modelers. Conventional employment numbers from the Bureau of Labor Statistics are not able to track such a new phenomenon because this economic ecosystem is so new. The research analyzed detailed information from The Conference Board Help-Wanted OnLine® (HWOL) database, a comprehensive and up-to-the-minute compilation of want ads, to estimate the number of jobs in the App Economy. The total number of App Economy jobs includes jobs at "pure" app firms such as Zynga, as well as app-related jobs at large companies such as Electronic Arts, Amazon, and AT&T, as well as app "infrastructure" jobs at core firms such as Google, Apple, and Facebook. In addition, the App Economy total includes employment spillovers to the rest of the economy.

2.28 Googling Apps

A new open-source HTML5 video player is available for download from Google. As well as being a good showcase app, it is also practically useful. It is the architectural core of the 60Minutesapp available in the Chrome Web Store. As well as being a basic video player, the app allows developers to create their own content. A category page also allows developers to build up a catalog of things they would like to watch. The user interface is fairly polished, and the developer can try it out at TheVideoPlayerSample. It can be customized, extended, or just used out of the box and populated with the developer's own content.

The app developer can modify the configuration via the config.json, which is a Java Script Object Notation (JSON) format file; using it, developers can customize the component for their own project. The code can be downloaded from GoogleCode (Figure 2.19). The project follows a model–view–controller (MVC) software design style architecture and uses the Closure JavaScript library; the final app can be compiled using Google's Closure compiler.

The Closure compiler is simply an optimizing JavaScript-to-JavaScript compiler, and so the end result is still an HTML5-compatible app. Some of the key features of the Google compiled apps are the following:

- The apps work with Chrome as well as other modern browsers
- Built-in support for sharing via Google+, Twitter, and Facebook
- The ability to subscribe to shows, watch episodes, and create play lists
- Support a video watching experience, including a full-screen view
- A categories page with an overview of the different shows available in the apps
- Notification of new episodes, when the apps are installed via Google's Chrome Web Store
- To ensure easy customization, all source files, including the Photoshop PSDs, are included
- Support for multiple video formats depending on what the user's browser supports, including WebM, Ogg, MP4, and even a Flash fallback

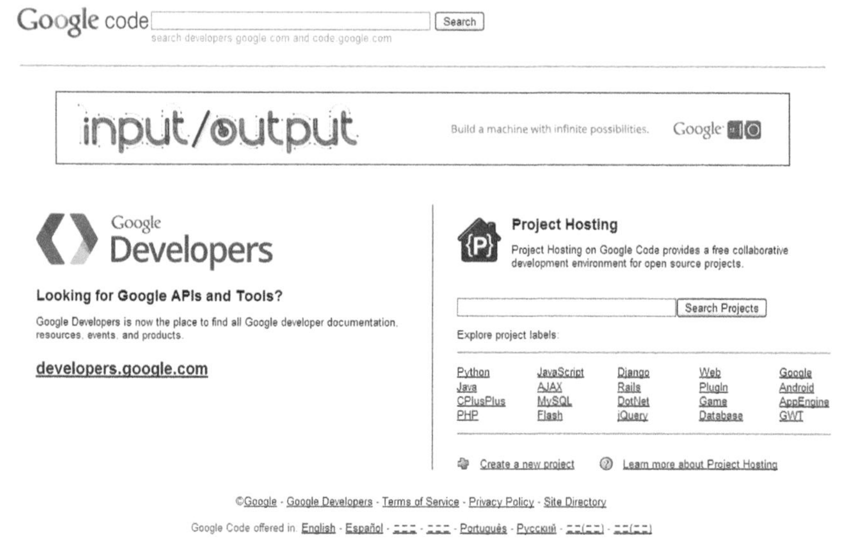

Figure 2.19 Google code site.

2.29 Apps Options

According to a recent survey by the Pew Internet Project (Pew.com), 43 percent of mobile users have apps on their devices, and most use them regularly. Business technology professionals surveyed by *InformationWeek* rank mobile apps as their highest enterprise software priority. So it is no wonder that most businesses, enterprises, brands, and marketers are looking to leverage the explosive growth of apps to help drive marketing, sales—and customer service. But building and deploying a quality app can be daunting. Most companies do not have deep mobile app development skills in-house. As previously mentioned, there are multiple operating systems to develop for and hundreds of OS-device combinations. There are three main ways to develop mobile apps:

1. Develop a native app for each of the dominant platforms: Android, Apple, BlackBerry, Microsoft, etc.
2. Buy and use a cross-platform development framework, leveraging its APIs to write code once but have the app run on multiple platforms.
3. Use a mobile enterprise app platform, which provides prebuilt, enterprise-ready integration with an enterprise's existing business systems via a vendor's framework, enabling rapid deployment of apps without much development effort or resources.

Let's start by looking at each app development option, as each has notable pros and cons. The main reason to develop a native app is because it needs access to a specific functionality provided by the mobile, such as an accelerometer, camera, or GPS, and also needs the benefits of integration with local processing information systems of that enterprise. While it is easy to understand the pros of a native app, there can be many more cons if the company does not have on-staff developers with considerable knowledge of the various mobile platforms. For this reason, most IT organizations will turn to outsiders to develop native apps.

First, the company or brand needs to make sure that the app vendor has experience with developing the specific type of app they want, not just experience with the operating system platform. So if a company is converting a customer relationship management (CRM) application for Apple mobiles, ensure that the app developer vendor has experience with multiple input forms and database-driven applications; otherwise there is a risk that the native app will have a clumsy interface.

Second, make sure the app developer vendor demonstrates the user interface before it builds the app and during development, using tools such as storyboards. Once it builds the full app, make sure that its quality assurance team is using handset-simulator software to test the native app on many different mobile devices. Different screen sizes, operating systems, processors, model devices, and RAM (Random Access Memory) can change the way an app functions.

Third, marketers and brands need to ensure that there is a service-level agreement governing how quickly the vendor developer will fix problems. App users will want a patch immediately; Android alone supports hundreds of devices and operating system configurations, and even Apple's closely controlled iOS has many versions that need testing.

The other major concern for developing native apps is security. While secure software development practices have been around for years, most mobile development groups simply are not following processes such as the Secure Software Development Life Cycle. Enterprises and companies can easily run into mobile app security problems. In past years, Citibank, Wells Fargo, and MasterCard have each released apps that stored data—including PINs and credit card numbers—insecurely on mobiles.

This type of vulnerability is well-documented within the Common Weakness Enumeration database (CWE-312). Compounding the security risk is the fact that native applications are custom-specific to their various operating systems—such as iOS and Android—which makes analyzing their security more difficult because many diagnostic tools that scan for vulnerabilities do not support these new proprietary platforms.

Since the introduction of Apple's App Store and Google's Android Market, app companies and app developers have seen success in marketing and selling their mobile applications to consumers. But with the adoption of mobiles in the enterprise, there is an equally, if not more lucrative opportunity to build mobile apps for businesses, brands, and marketers. Enterprise IT departments simply cannot keep up with the ever-growing number of mobile technologies, platforms, and devices, and will need the independent mobile app developer's assistance.

Partnerpedia (Partnerpedia.com) (<http://www.partnerpedia.com/>), a mobile app management and marketplace portal, recently conducted a survey that showed great interest among app developers to offer mobile apps to businesses, marketers, and brands. Over 80 percent of the 200+ app developer respondents indicated that they plan to offer apps to business and marketing customers. The prime reason for this is that the consumer market is overcrowded and there are higher margins in business and marketing apps.

Android led the way when it came to targeted mobile platforms, at 82.5 percent of respondents. Apple's iOS was not far behind, with nearly 78 percent. Android's openness could be the key driver for success in the enterprise; it makes it easier for app developers to build and sell enterprise and marketing mobile apps. One challenge app developers will face is deciding which operating system to target and whether to build native apps or HTML5 multi-platform apps.

There are tools and frameworks out there that allow developers to build both native and HTML5 apps that run on many operating systems, including Appcelerator® (<http://www.appcelerator.com/>), Adobe PhoneGap™ (<http://phonegap.com/>), and Netbiscuits (<http://www.netbiscuits.com/>). Another challenge is Android itself, as app developers are working with various Android versions such as Honeycomb or Gingerbread; whereas for iOS, most are developing on the latest version of Apple's mobile operating system.

The need for mobile apps in the enterprise is going to drive business and marketing innovations by creating a more competitive market, mirroring what has taken place in the consumer app market. Opening up traditional IT to include third-party app developers is going to revolutionize the business software market and will ultimately increase employee, marketing and company performance.

Mobile app development for business and marketing is the future; however, there are many companies, ad networks, mobile sites, and app developers to partner up with. So before making a decision, take the time to consider several issues: what type of app to create (i.e., game, social networking, utility, etc.); what is the target goal of the app (i.e., is it to increase downloads, generate user engagement, increase sales, or generate buzz); and perhaps most importantly, create consumer behavioral data for mobile analytics.

2.30 App Security

With every business, from the tiniest company to the largest enterprise looking to plant its flag in the ground with regard to mobile applications, the mobile app development boom is on in a very big way. Amid this blind rush to beat the competition to market, mobile developers are feeling their way around in the dark, and with a development environment still in its infancy and no real standards to lead the way, it is an adventure for all parties involved.

Particularly scary to many security professionals is the fact that the speedy mobile app development cycle and this lack of experience in the platforms is causing coders to throw all of those secure development principles the industry has fought for over the past five years right out the window when it comes to mobile apps. Rapid and agile development causes changes to happen in very short iterations; thus, security gets overlooked and becomes a nice thing to do but rarely gets done. This occurs even in large corporations, such as incidents and breaches with Google Wallet™ and, even worse with app development start-ups.

When TechCrunch (<http://techcrunch.com/>) announces the hottest new start-up of the day, week, or month, almost every single one of those companies lack secure coding practices and are rarely even concerned until something goes wrong. Most of the time, they are not even aware of these issues; even the big mobile platform vendors such as Apple and Google have only just now started to think about secure mobile coding and have mainly been interested in looking the other way. The difficulty is that even for established firms that are aware of their risks and want to securely code their apps, there are few standards for development and very few tools for testing code for vulnerabilities.

Apps testing methodologies are quite a bit different from testing normal applications. Identifying the key risks, the lack of standards, and the technologies needed to test them properly are a major challenge. As a result, apps are already starting to flood the market with critical vulnerabilities that put customers, brands, and business resources at risk. For example, apps developers are not testing the mobile services that they are using in the cloud and are introducing a whole spate of encryption flaws through their apps, such as leaving unencrypted passwords in data cache files.

In fact, viaForensics, a digital forensics and security firm, reported that 76 percent of popular consumer applications running on Android and Apple mobiles stored passwords in plaintext. In other words, apps are storing too much data on mobiles in a non-encrypted format—where there are all sorts of passwords and other information (e.g., social security numbers and credit card information) being exposed. However, the Open Web Application Security Project (OWASP; <https://www.owasp.org/index.php/Main_Page>) has been working on mobile app security via their OWASP's Mobile Security Project (OWASP.org), which aims to offer app developers and marketers the tools and resources for writing and supporting secure apps. The OWASP includes a threat model, training, and platform-specific guidelines.

But meanwhile, mobile app vulnerabilities are showing signs of growing pains. Google Wallet, for example, was shown in a different viaForensics report to be storing all sensitive information, except for credit cards, locally on the devices in plaintext, where hackers can easily crack the PIN at Google Wallet on rooted mobiles. The lesson for app developers is that as organizations release apps that tap into sensitive information and tap into payment systems such as Google Wallet, they need to be mindful of the inherent risks. Brands and businesses need to test both the client and services portions of the app, using a combination of both dynamic and static testing technology and both internal and external test teams.

How safe is data stored on mobiles? Not very; in fact, 76 percent of all popular consumer apps running on Android and Apple mobiles store usernames as plaintext, and 10 percent, including Hushmail, LinkedIn, and Skype, also store passwords as plaintext. This is according to a recently released report from viaForensics (OWASP.org). For the study, company researchers evaluated 100 popular consumer apps that run on Android, as well as Apple's iOS operating system. The firm's application assessments found that numerous apps store data, including usernames, as plaintext on mobiles. Many systems require only username and password, so having the username means that 50 percent of the puzzle is solved. In addition, people often reuse their usernames so it will generally work on many online services.

Arguably worse, however, is when apps fail to encrypt even more sensitive information, such as passwords, which poses a risk to consumers because mobiles are frequently lost or transferred, and because malware could potentially grab the data. Some risks, such as stored passwords or credit card numbers, are clearly greater than others. When it comes to the security of mobile consumer apps, the tested social networking apps fared the worst, with 74 percent earning a fail. This places the user at a significant increased risk for identity or financial theft. Other application categories fared better, including productivity apps (43 percent failed), mobile financial apps (25 percent failed), and retail apps (14 percent failed). While the retail application failure rate looks low, no retail applications actually passed the test.

On both Android and iOS, apps that store sensitive data insecurely as previously mentioned include Hushmail, LinkedIn, Skype, and WordPress. Meanwhile, on Android alone, apps that store sensitive data insecurely include Android Mail, for Exchange and Hotmail, Gmail, Netflix, and Yahoo! Mail. Apple mobiles apps that store sensitive data insecurely include Chase for banking and iPhone Mail for Exchange and Gmail. Numerous other applications, however, also store nonsensitive data in unencrypted format, including mobile software from Amazon.com, Best Buy, Facebook, and Twitter.

Of course, all the above apps rely at least in part on the underlying operating system to remain secure. Accordingly, which is more secure? Users of Apple mobiles appear to have better out-of-the-box protection, according to viaForensics (OWASP.org). According to the security firm, Apple has made more efforts toward data protection via their iOS platform, compared to Android. However, users still face risks due to malware that can compromise the mobile, or data recovery from lost/stolen devices.

That said, changes are afoot. Google released Android version 3.0, aka Honeycomb; notably, the operating system upgrade will encrypt the user partition on Android mobiles—but so far, it is only available for tablets, not phones. Therefore, if someone acquires a lost or stolen phone, or a malware program, they can gain root access on an Android mobile where they then have full access to the user partition and its data. However, Apple's iOS is not bulletproof, or standing still either. Apple upgraded its mobile operating system with better encryption as of the 4.0 version, but already forensics researchers and toolmakers have cracked the iOS data security scheme and released automated tools that can recover much of the information stored by the iOS mobiles, providing they can crack the device's password.

In other words, the security of an iOS device is very much up to its owner. That is, if the mobile user does not activate the data protection feature by setting a passcode, the files are not fully protected. Furthermore, various tools exist to uncover the user's passcode with varying degrees of success, depending on the strength of passcode used.

Presently, Apple iOS and RIM BlackBerry mobiles offer levels of security beyond what is available on Android, simply because of the extent to which Apple and RIM control their mobile operating system environments. What the app developer and marketer need to be aware of are the security dangers they face with these various operating systems and the numerous mobiles currently in the marketplace. The lesson here is to ensure that adequate security testing is required before releasing any consumer apps.

2.31 Local Company Apps

Small businesses that use mobile apps to help manage their operations are saving more than 370 million of their own hours and over 725 million employee hours annually, according to a study by the Small Business and Entrepreneur Council (SBE Council.org). The study showed that 31 percent of small firms with twenty or fewer employees were using mobile apps. Of those using apps for their business, owners estimated that they saved an average of 5.6 hours per week. In conjunction with the total number of small firms in the United States, the study estimated that 1.28 million small-business owners are saving time with mobile app usage.

This translates into savings of hours of paper-pushing, administrative work, customer research, extra driving trips, and unproductive downtime, the SBE reports. Mobile apps are allowing small-business owners and their employees to get more time out of their day and to focus on higher-value work. Based on the amount of employee-hours saved and the average pay for small-business employees, the study estimates that small firms are saving $275 per week, adding up to an average of over $14,000 per year.

The use of mobile apps, which can spare firms the time spent on monotonous tasks such as bookkeeping, document sharing, and travel, can save small-business owners an average of $377 per month, adding up to over $4,000 per year. The SBE study emphasizes that such estimates are conservative, as only firms with twenty or fewer employers were surveyed.

As a result of app usage, 49 percent of the owners surveyed said their firm was able to spend more time on increasing sales and creating new revenue streams, while 36 percent said app usage allowed them to cut overhead costs. Perhaps most significant, 51 percent of small-firm owners said app usage allowed their business to become more competitive and better able to maneuver through an uncertain economy.

Nowadays, an app can allow a company to share spreadsheets and process credit card payments. The mobile app revolution makes it easy and inexpensive—or even free—for entrepreneurs to manage their entire operation with or without an office. Entrepreneurs are increasingly turning to the more than 1 million mobile apps available to meet their business needs. Forrester Research estimates that revenue from customers downloading mobile apps will reach $38 billion by 2015 (Forrester.com).

Apps can save time and money; boost sales and productivity, and help small-business owners improve efficiency. They can also give consumers new ways to interact and communicate with companies and brands. Mobile apps save small businesses 725 million hours per year, for an estimated $17.6 billion in savings, according to a survey by the SBE Council (SBE Council.org). More than one-third of U.S. small-business owners surveyed by Intuit (Intuit.com) said annual business growth was the biggest worry, but nearly half said it is important to be able to run their business on mobiles. Here are some examples of how entrepreneurs are using mobile apps.

Pinterest (<http://pinterest.com/>) is a popular e-pinboard for self-promoting a company; some savvy small-business owners use it to showcase their products or services. The app can be used by small businesses to help promote and expand their companies. Pinterest can be used to "pin" and share favorite items on the Web. The app can also be used to create "vision boards" of ideas and shows them to clients on tablets and mobiles. If people like what they see, they will re-pin it. The app can also be used to drive traffic to their website and works best for businesses with a visual side. For example, an interior designer could use it to pin photos of redecorated rooms or a landscaper can share pruning advice. It is an invitation-only site but people can follow each other as they do on Twitter.

Then there is Dropbox (<https://www.dropbox.com/>), an app and a free service that lets a business share its photos, docs, and videos anywhere, and share them easily with their potential new clients. It prevents having to email documents back and forth. Remote document access apps are the third most popular (used by 41 percent) among small businesses, according to the Small Business & Entrepreneurship Council (SBE Council.org) survey.

Next in popularity are mobile payment apps, which are designed for anyone who sells products and services on-the-go. They can also cut payment costs. Food vendors, photographers, hairstylists, and even bands use these apps. A variety of apps are available with small card swipes of credit cards, like those of Square, most of which are free and fit into the audio jack of a mobile device. Authorization of a credit or debit card is done as the payment is processed; customers sign in with a finger. The following are the most popular small-business apps:

- Dropbox:
  - Cost: Free up to 2 gigabytes of storage
  - Mobiles: Apple, Android, BlackBerry
  - Users can take documents, photos, and videos anywhere and share them. Dropbox Pro ($19.99 a month) upgrades the free 2 GB of storage to 100 GB and provides access to a service copyright agent.
- Evernote:
  - Cost: Free
  - Mobiles: Apple, Android
  - Users can take notes, record phone calls or reminder notes, capture photos, create to-do lists, and share data. Evernote also categorizes and tags items. It syncs with Facebook and Twitter. Users can pay more for more features, such as a lock system and extra storage.
- Expensify:
  - Cost: Free to download (basic version)
  - Mobiles: Apple, Android, BlackBerry, Windows Phone, Palm
  - It automates the entire expense report process. It links to a credit card, uploads receipts, reimburses via PayPal, and syncs with Evernote. Most purchases under $75 are eligible for automatic e-receipts. There is a pay option for more receipt scans after the first free ten scans.
- IntuitGoPayment:
  - Cost: Free to download with free card reader
  - Mobiles: Apple, Android
  - A pay-as-you-go version charges 2.7 percent per swipe or 3.7 percent per key-in. High-volume users can pay a monthly fee of $12.95 plus 1.7 percent to 2.7 percent per transaction. Issuer Intuit recently released a newly designed, smaller card reader.
- OfficeTime:
  - Cost: Free
  - Mobiles: Apple
  - It is designed for freelancers, consultants, lawyers, and others who bill by the hour. The app, which links to Excel, tracks billable time, projects, and expenses.
- Pinterest:
  - Cost: Free
  - Mobiles: Apple
  - It's a virtual pinboard to save and share products and services. Businesses such as architects and bakers can showcase their work. If it's liked by others, it will be shared repeatedly.

In 2016, total global mobile application revenue will reach an estimated $46 billion, according to ABI Research. That figure, the research firm says, includes pay-per-download, in-app purchases, subscriptions, and advertising. Last year (2011), mobile app revenue hit $8.5 billion, ABI research says. This year, in-app purchases will likely outpace pay-per-download revenue, but according to ABI Research (ABI Research.com), the number of people buying content in apps will not grow all that much, potentially putting that revenue opportunity's future in doubt—however, mobile apps are a big business.

As a revenue model, in-app purchase is very limited today; with the vast majority of current in-app revenue being generated by a tiny percentage of people who are highly committed mobile game players. In the world of business, few things are more valuable than customer loyalty. And it is a phenomenon that is distinctly contagious. If you have customers who cannot get enough of the gourmet burgers that your restaurant serves, or clients who swear by your company's automotive products, they are certainly going to tell their friends.

Throw some consumer incentives into the mix and you may have hooked some customers for life.

Enter Punchcard (<http://www.punchcard.com/>), a company created through the Pasadena-based business incubator, Idealab. Punchcard (Figure 2.20) has created a mobile customer-loyalty app that not only benefits customers with discounts, but also provides merchants with valuable information about their customers' buying habits. The app is simple to use; all a customer has to do is take a picture of their receipt at the end of each transaction. Then they select the business where they bought the product and earn punches on a punchcard. The Punchcard app replaces physical punchcards that businesses often have. With Punchcard, a merchant can determine a customer's buying habits.

Punchcard allows merchants to segment out their customers based on what they have bought. As a result, businesses can offer specific kinds of incentives, such as a "buy-two, get-one-free" deal on products they know a consumer will want. Consumers like Punchcard because they get cash back, and merchants like it because it helps them acquire customers more efficiently. This app is the perfect tool for aggregating consumer behaviors and data mining mobile devices and their owners.

The app is free for consumers. Merchants pay either $29 or $99 a month per business location to participate. The $29 package is essentially an electronic version of the old punchcard system. The $99 version offers more in-depth information regarding each customer's buying habits. Punchcard really helps small businesses keep track of the loyal customers; it allows small firms and their employees to get to know the customers on a first-name basis.

2.32 The $ Apps

A good example of killer app design and strategy for marketing and generating revenue are the price comparison offerings from eBay and Amazon; the goals of

Figure 2.20 Punchcard offers a huge insight into a business customer base.

both are mutually beneficial to consumers and the enterprises that created them. The benefit to consumers is finding the cheapest price anywhere in the globe—the benefit to eBay and Amazon is increased sales. But more important is the aggregation of a tremendous amount of shopping and pricing data that is ripe for mobile analytics for segmenting and modeling the behaviors and presence of different consumers, across different countries, via multiple product categories.

The eBayMobile app (<http://mobile.ebay.com/>) is an example of great functionality with its universal price comparison feature. Such transparent capabilities create enduring customer loyalty and ensure their engagement in the future. Amazon is also giving comparison shoppers a quick-and-easy way to get discounts at its mobile site. The Amazon Price Check app (http://www.amazon.com/gp/feature.html?ie=UTF8&docId=aw_ppricecheck_iphone_mobile; Figure 2.21) is free and works on all major mobiles. The app can scan a product's barcode, take a picture of the product, or the user can say or type its name. The app then provides the Amazon price—and can put it in its Amazon's online shopping cart with credit card checkout capability; discounts are provided on some time-sensitive purchases.)

Both of these price comparison apps are not just a one-stop destination, but are, in fact, a starting point for search shopping, pricing functionality, and social sharing with friends via consumers' mobiles. Because they are transparently independent, they are endearing to mobile users, creating a tremendous amount of consumer loyalty. For their creators, they are a conduit to a massive amount of behavioral consumer and pricing data for mobile analytics.

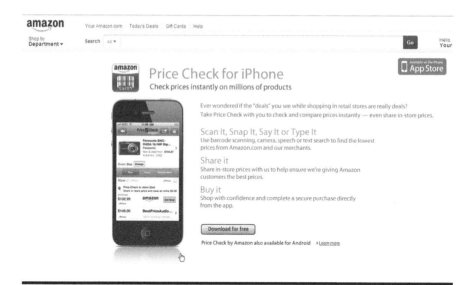

Figure 2.21   The Amazon Price Check app.

Another app that turns mobiles into a cash register, as previously mentioned, and can be used to track the history of customer purchases is Square (Figure 2.22). This analytics program lets small businesses "slice and dice" sales data. With over 75 percent of U.S. merchants intending to buy a tablet over the next year; according to the National Retail Federation (NFR.com), this type of app is ideal for merchants to segment their customer mobile purchase behaviors.

Small merchants are often saddled with clunky POS systems that cost thousands of dollars a year to maintain, and take hours to install. However, they can download the Square Register app (<https://squareup.com/register>) for free and install their inventory in less than an hour. Square makes its money by collecting small transaction fees. The original Square card reader, a small white plastic device for accepting payments, has more than a million users nationwide and works on Apple and Android mobiles.

2.33 Ambient Social Network Apps

People discovery apps are starting up; for example, Highlight (<http://highlight/about.html>) is an Apple app designed to reveal real-life connections a user did not know they had, as well as alert them to the presence of friends they might otherwise miss. Highlight works by rummaging through a user's Facebook account to see who they know and what topics they like. Then it uses the mobile's GPS to inform them when a friend is in their vicinity. Highlight calls itself a "crowd

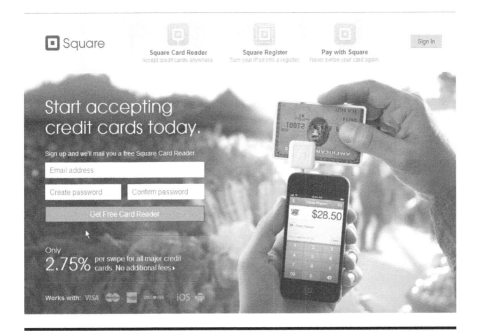

Figure 2.22   The Square app card reader.

sourced recommendation" app that lets users find the very best places around the world. Whether they are trying to discover the nicest bar in a town or looking to plan their next vacation, Highlight uses its advanced recommendation engine incorporating data from shopping, picture, and music sites, such as Foursquare, Flickr, Yelp, and Last.fm.

Highlight monitors the whereabouts of mobiles continuously and automatically, and shares them with fellow members, both in and outside the user's existing circle of friends. Other social apps include Glancee (<http://www.glancee. com/>)—recently acquired by Facebook—which pinpoints others who are "steps away" and shares the user's interests. Another is OkCupid (<http://www.okcupid. com/>), which locates nearby singles by compatibility.

Then there is Pair by TenthBit Inc. (http://trypair.com/), an app for sending a "thumbkiss" between two users who slide their thumbs over the screen of their mobiles, which when they line up, the screen vibrates and the screen turns red —a pretty interesting idea for an app that couples in long-distance relationships will no doubt enjoy (Figure 2.23). Once two partners download the app, they can send text messages, photos, and videos to each other without worrying about interlopers viewing their intimate exchanges. Pair also lets user draw together, and have a joint to-do list. Pair is the logical extension of a trend toward privacy in app social networks.

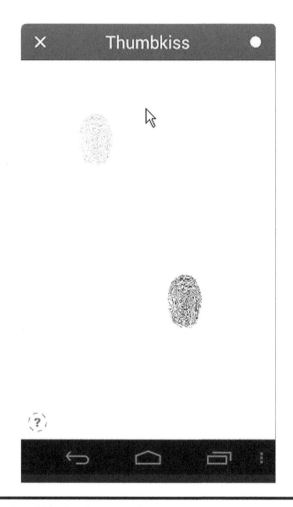

Figure 2.23 Thumbkiss by the app Pair.

2.33 App Developers

As previously mentioned, one of the options for marketers and brands is to outsource the development of their app to experienced professional developers. The following are some of the best-of-breed in this rapidly evolving new market sector:

- Elance® (<https://www.elance.com/>): This is an IT skill exchange site with over 300,000 freelancers. Mobile marketers and brands can post job bids to have custom-developed apps. A recent skill search for "mobile app professionals" found over 40,000 available contractors and firms.
- MyFirstMobileApp (<http://www.myfirstmobileapp.com/>): They are app developers for all the major mobiles. They offer full-spectrum, end-to-end

services across diverse categories such as business, utilities, entertainment, and gaming among various app categories. They offer custom apps development on all types of mobile platforms, including Apple iOS, Google Android, Windows Mobile and Phone, BlackBerry OS, and Symbian OS.

■ Zco Corporation (<http://www.zco.com/>): Mobile and enterprise custom app development service extends to all major mobile platforms. They also provide 3-D animation and digital marketing. Clients provide the concept they have in mind for their apps and Zco will build it. They offer app development for multi-platform and back-end system integration capabilities.

■ Mutual Mobile™ (<http://www.mutualmobile.com/>): This is the largest mobile app development company in the United States. They assist clients in taking full advantage of mobiles. They provide expertise and the ability to execute the app design, engineering, and project management resources for all major mobiles.

■ Bianor: They provide free analysis and rapid development, and specialize in complete mobile solutions ranging from telecom-grade back-end services to app development.

■ XCubeLabs: Experienced, low-cost, fast results, and free quotes, their specialization includes developing apps across multiple verticals. They have developed over 400 apps and have strong technical competency, well-defined methodology, and a team of expert engineers and designers.

■ BestFitMobile: They support all major mobile platforms. They design, develop, and optimize mobile solutions for companies that want to engage their customers.

2.34 App Demographics

Both Apple and Android apps developed by other companies capture different types of important demographic data that can be used for clustering and segmentation via mobile analytics. These apps can be used to better track and target consumers and their mobiles with relevant content, products, and services based on such parameters as age, gender, location, and interests. Being relevant in the marketing to mobiles is of paramount importance to brands, enterprises, developers, analysts, and marketers.

Android apps that capture age and gender are limited to Pandora® and MyspaceMobile; however, the following Android apps do capture location: Alchemy, BarcodeScanner, BeautifulWidgets, CalorieCounter, CardioTrainer, CBSNews, Foursquare, FoxNews, Groupon, MoviesbyFlixster, Pandora, PaperToss, Shazam, TheCouponApp, Tossit, TweetCaster, USYellowPages, Weather&ToggleWidget, TheWeatherChannel, and WeatherBug.

Apple captures age and gender on Grindr and Pandora; the Apple apps that capture location are AngryBirds, CBSNews, Dictionary.com, MyFitnessPal, Ninjump, TheMoronTest, TheWeatherChannel, and Grub.

Figure 2.24 An app developer SDK.

App developers can insert an SDK from an advertising network, such as Greystripe (Figure 2.24). That is a common practice among app developers, who use these ready-made kits to place ads and generate revenue by targeting devices based on their interests and/or location, as well as aggregating important user demographics for mobile analytics.

2.35 App Triangulating Services

There are also several mobile marketing and triangulating firms that developers, publishers, and marketers can leverage to track when and where their apps go and are located in real-time.

- App Annie (<http://www.appannie.com/>): More than 75,000 publishers and developers of Apple apps use this service to track downloads, sales, and reviews for free.
- MobileDefense™ (<https://www.mobiledefense.com/>): Originally developed to track lost or stolen digital devices, the firm is now marketing to them. They were the first security tracking solution for Android, and provide military software to mobile marketers.
- Glympse® (<http://www.glympse.com/>): Offers software that allows GPS-enabled mobile devices to be tracked and is an easy way for mobiles to safely

Figure 2.25 Triangulation system by Navizon. (*Source:* From http://www.navizon.com/.)

share their location in real-time to anyone via email, SMS, Facebook, or Twitter.

■ MobiWee (<http://www.mobiwee.com/>): Runs on Windows, Android, Apple, and BlackBerry mobiles. The software offers one-click remote access and transfers files from any stationary computer to mobiles and back, for contact backup, restore, merge ,and transfer access.

■ Navizon (<http://www.navizon.com/>): Offers an SDK for app tracking purposes. Navizon (Figure 2.25) can provide the location, latitude, and longitude of mobiles based on the triangulation of Wi-Fi or cellular signals. Their Client-Side SDK is a library that developers, publishers, and marketers can include in their app. The Navizon positioning system can collect the triangulated information surrounding mobiles and convert it into a geographic location, which is then relayed to clients to use any way they see fit. They also offer an indoor tracking triangulation system as well as advance mobile analytics.

2.36 App Ad Agencies

There are also digital ad agencies that are assisting their clients and brands with mobile sites and app ad campaigns. Increasingly, ad campaigns are going mobile; it is the new frontier in the world of advertising and marketing.

- DataXu (<http://www.dataxu.com/>): Provides media management platform for ad campaigns across all channels. The agency starts by assimilating the client's existing data. Then it analyzes it, looking for patterns of behavior that help it understand or differentiate levels of intent among their target audience. Their DX Mobile is a demand side platform for mobile advertisers.

- BBDO (<http://www.bbdo.com/>): Offers marketers insight into the everyday routines of people across the globe via several channels, including mobile. BBDO is a worldwide advertising agency network, with headquarters in New York.

- Razorfish (<http://www.razorfish.com/>): A large, interactive agency offering digital advertising, helping companies build their brand through strategy planning, interactive design, social influence marketing, search and email marketing, analytics, technology architecture, and development.

- GeniusRocket (<http://www.geniusrocket.com/>): An ad agency that connects creative talent to marketers, publishers, companies, and their brand managers; they have over 200 production developers, 100 animation teams, 50 copywriters, and 50 creative directors.

- Lotame (<http://www.lotame.com/>): This agency provides social media ad campaigns via their Crowd Control Technology, which seamlessly targets consumers across multiple media outlets, including mobiles. Their affinity reports identify related audience attributes and target segments with the highest likelihood to perform desired behaviors, such as interacting with a brand.

- MediaMath (<http://www.mediamath.com/>): They provide digital media services via their demand-side platform (DSP). This agency uses data to understand consumer behavior and identify opportunities. They translate those insights into integrated marketing strategies in the mobile channel. They use advanced analytics to identify the media and audiences that will best reach those goals, and quantify the value of each.

- Big Spaceship (<http://www.bigspaceship.com/terms-of-service/>): This digital creative agency specializes in Web design, strategy, marketing, and branding. They are designers who work with brands to solve their business problems.

- [x+1] (<http://www.xplusone.com/about/company-overview/>): This agency provides audience targeting via digital media. [x+1] works with brands, agencies, and media companies to determine the most valuable customer attributes. Their Predictive Optimization Engine enables automated, real-time decision-making and personalization.

2.37 App Analytics

Because the main purpose for creating an app is to generate device data for leveraging mobile analytics by enterprises, brands, and marketers, engaging such a metric

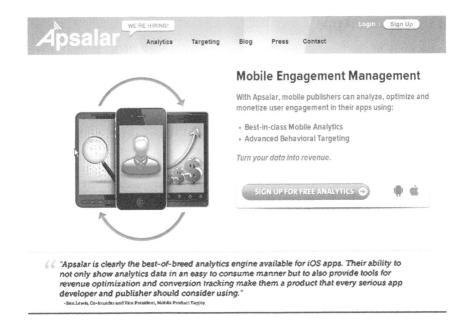

Figure 2.26   Mobile app analytic tool Apsalar.

reporting company as Apsalar (<http://apsalar.com/>) is critical. The start-up offers its Engagement Index and Revenue Analytics tools that give mobile app publishers unprecedented levels of information to quantify and optimize user engagement and the monetization of their apps. Apsalar can report on app activity via its Mobile Engagement Management (MEM) platform (Figure 2.26), and provide mobile analytics and behavioral targeting solutions for iOS and Android apps via its free ApScience mobile analytics service.

The Apsalar ApScience service provides mobile publishers with the capability to better understand the level of engagement and revenue generated from their apps. The service allows app developers to assign engagement or revenue values to select user actions and events. Engagement values are determined by the publisher in accordance with whatever engagement scale they choose to follow, whereas revenue is determined by the actual amount of money collected from users of the app.

These ApScience metrics are tracked for each unique user when they hit any event that has either an engagement or revenue value associated with it. The values are then used to provide information in multiple reports and analyses so that publishers and developers can better understand overall engagement and the average engagement per user (AEPU) of their app, as well as overall revenue and average revenue per user (ARPU). With the Revenue Analytics and Engagement Index, mobile app developers can measure how changes to their apps impact engagement and revenue. For example, an app acquisition campaign can track how the revenue from their customers evolves on a daily basis after the download.

2.38 Ads as Apps

A new study by the Institute for Communication Research (ICR) at Illinois State University (Illinoisstate.edu) has concluded that interactive apps for mobiles may constitute one of the most powerful advertising tools developed to date. The study confirmed that using branded mobile apps increases a consumer's general interest in product categories and improves the attitude that the consumer may have toward the sponsoring brand. The researchers also found that apps that are informational in nature or utilitarian—such as the price comparison apps—were more likely to engage users than those where the app focused on entertainment or gaming.

Apps can be a way for advertisers and marketers to reach across traditional product or gender boundaries to appeal to new types of consumers. Unlike viewing information through a print ad, a media spot, or on a website, consumers will process a company's messages more deeply if they do so using an app that they have decided to download to their mobile device. The study concluded that because consumers have a more personal connection with their mobiles than with a website, they have a deeper level of interactivity.

Another benefit of the ads as apps is that it goes with the consumer once they download it, becoming their companion whenever and wherever they go. The very personal nature of mobiles is that they are practically extensions of their owners; the ICR study concluded that advertisers need to adopt new rules of conversation with mobile users. The researchers used eight branded apps; half of the brands were in product categories that predominantly target men and the other half targeted women. The four male-targeted brands were Best Buy, Gillette®, BMW, and Weber, while the four female-targeted brands were Gap, Kraft, Lancome, and Target.

2.39 The Future of Apps

App marketing is part of the future of advertising, with more major brands turning to mobiles to target their increasingly fragmented consumers. The following checklist details some issues that app developers and marketers need to focus on in a continuously evolving manner that is engaging and productive:

1. Identify app areas of opportunity—and take action.
2. Validate the accuracy of the app data—continuously.
3. Set app team goals and incentives—make it a nonstop process.
4. Focus on actionable consumer data—at their mobile site and app.
5. Set success app metrics—continuously measure key performance indicators (KPIs).
6. Monetize all app models—measure ROI, sales, conversions, downloads, profitability, revenue, and WOM.

A recent Gallup poll found that nearly half of all mobile owners who download and use apps said that they do not mind mobile analytics as long as the apps have their permission.

Chapter 3

Mobile Data

3.1 How It Works

For the data mining of mobiles to take place, the importance of knowing how mobile data is created and can be aggregated for analytics is paramount. So what makes current mobiles more like computers is their ability to connect to the Internet and transfer data, while dictating their location and consumer interests. For a mobile digital device, it has been a long and difficult road from 1G to 4G. To date mobiles have gone through roughly three generations, as can be seen in Figure 3.1, with the fourth generation now being introduced.

Figure 3.1 shows the rough development of mobile phone technology, with the newest evolution being 4G; the whole concept is to increase and speed up the delivery of mobile data. The first generation of mobiles used a mostly forgotten analog technology that served to get the mobile market off the ground. Most of the mobiles in use at the moment are in the 3G evolution of the original 2G design based on a technology called Global System for Mobile (GSM) communications.

In Europe, people tend to think that mobiles are synonymous with GSM, but this is not quite true. In the early days, there were a number of different and incompatible systems deployed worldwide. Even today, GSM only accounts for just over 70 percent of the world market, with alternatives such as the code division multiple access (CDMA), which is a channel access method used by various radio communications technologies.

Then there is the Integrated Digital Enhanced Network (iDEN), which is a mobile telecommunications technology developed by Motorola that provides its users with the benefits of a trunked radio and a cellular telephone. iDEN places more users in a given spectral space, compared to analog cellular and two-way radio systems, by using speech compression and time division multiple access (TDMA),

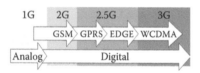

Figure 3.1   The progression of mobile data. (From GSMA.com.)

which is a channel access method for shared medium networks. It allows several users to share the same frequency channel by dividing the signal into different time slots. The users transmit in rapid succession, one after the other, each using their own time slot. TDMA currently makes up about 8 percent of the current data mobile data market.

Even so, it is fair to say that GSM is the global standard for roaming and in Europe is the only standard. Although 3G is very definitely the desirable goal, there are some interim measures that improve on the data transmission capabilities of 2G so much that it is fair to call the result 2.5G. In theory, the progress from 2G, through 2.5G to 3G should be smooth and backward compatible—as long as it is based on GSM-based systems. As with 2G mobile phone technology, there is a range of different and incompatible 3G developments. For data mining mobiles, it is important to know how mobile data is created and the features and intricacies it provides for advanced consumer behavioral analytics.

To distinguish these from the GSM-compatible versions, the term "3GSM" is sometimes used. It is worth knowing that the jargon used varies greatly and does nothing to help work out what is compatible with what. 3GSM is essentially a GSM network that uses Wideband-CDMA (W-CDMA) as its transmission method. It is currently claimed that 85 percent of the world's network operators have chosen 3GSM (3gpp.org) as their underlying technology, but it is important to be aware that there are variations and alternative technologies that are also referred to as 3G. For example, although Japan has implemented a system based on W-CDMA, it uses its own special variation. When it comes to 4G, the situation is even more confusing and disparate.

It is time to look at the details of how it works and, in particular, what makes 3G so much better than 2G. When the 2G digital networks were being implemented, the state of radio communications was nowhere near as advanced as it is today. The system used depended on a mix of traditional wireless communications and computer techniques. A block of frequencies is allocated enough to provide around 800 one-way signal channels; and for two-way or full duplex communication, you need two channels per phone, bringing this down to 400 connections. This clearly is not enough to offer communications channels to a very large population, and the key to making things work is frequency reuse.

By using low power transmission, a block of frequencies can be used to provide, say, 100 duplex channels over a small area—called a cell. A hundred channels can also be provided in each of the neighboring cells without interference by using a

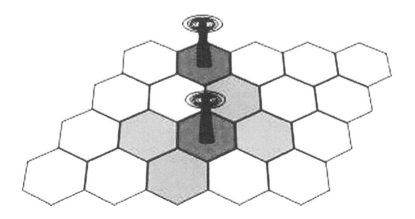

Figure 3.2   The triangulation of mobiles.

different set of frequencies. Once a cell is far enough away from the first cell, the same set of frequencies can be used over again, as in the case of the two dark cells in Figure 3.2.

The cell principle allows the same set of frequencies to be used over and over again without interference. The size of a cell varies greatly, depending on population and usage. Today there is a move toward the microcell that serves maybe a building or a block. Using microcells or even smaller cell sizes is one way to increase the overall data-carrying capacity of a set of frequencies. The cell principle is the reason why we sometimes call GSM devices "cell phones." However, there is more to GSM than just the use of cells.

Allocating a single physical channel to a single connection would be a huge waste of resources. Instead, each physical channel is divided into eight time slots (0.577 milliseconds in duration), which are time-shared between as many as eight users. This is called TDMA, and this plus the use of different frequencies for each connection or what is commonly known as Frequency Division Multiple Access (FDMA) are the two main technologies that make up GSM.

A cell's allocation of frequencies is split into two portions used for up and down links. A duplex (two-way) connection uses one frequency band from the uplink and one from the downlink set separated by 45 MHz. Each 200-KHz wide channel is further divided into eight time slots (Figure 3.3). As 2G is a digital network, one would expect little trouble in making a digital connection; but given that the technology was designed and optimized for voice communication, it is not quite that simple.

The most basic way is to use a "modem" built into the mobile to dial up another modem and make a Circuit Switched Data (CSD) connection. In theory, a single time slot should be capable of carrying 34 Kbits/s, but the need for error correction and encryption reduces this by at least a third, thus producing a usable data rate of around 9.6 Kbits/s. It also takes longer for a call to be set up because the cell

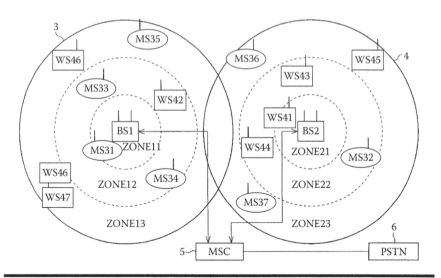

Figure 3.3   The configuration of channels via slots.

handles a data call differently. Instead of routing the call in the usual way, it actually dials the number that you supplied and connects to your ISP's (Internet Service Provider's) modem. This also adds a latency of up to 1 second.

An improved CSD connection, High-Speed Circuit Switched Data (HSCSD), is offered by some operators. This works by using multiple, up to eight, TDMA slots per user and can achieve, taking count of some inefficiencies in using multiple slots, up to 57.6 Kbits/s. In practice, the operator limits the number of slots that can be used so as to allow other users to make calls. Users also need a mobile that supports HSCSD. CSD and HSCSD are both point-to-point connections, which makes sense for dialing a private network but not for connecting to the Internet. GPRS (General Packet Radio Service) is a direct connection to the Internet. The mobile sends and receives IP (Internet Protocol) data packets via the cell to the Internet. In this sense, the mobile provider company is acting as an ISP. Unlike CSD, there is no setup, and data packets can be sent when required and received at any time. That is why GPRS is an "always-on" service.

As in the case of HSCSD, GPRS can also use multiple slots to send a packet and, with improved coding, also increases the useful data that can be set in a slot from 9.6 to 13.4 Kbits/s to give a maximum data rate of approximately 100 Kbits/s. In practice, due to limitations in cell hardware, GPRS uses a maximum of four slots in the downlink direction, giving a transfer rate of around 40 Kbits/s and as few as one slot for uplink providing around 10 Kbits/s. GPRS is ideal for burst data transfer, whereas HSCSD is better for bulk connections. It is also worth noting that HSCSD is generally charged by connection time, whereas GPRS is charged by amount of data transferred. Some operators also prioritize voice traffic so that GPRS is only allocated time slots when they would otherwise be unused.

Then there is EDGE (Enhanced Data rates for GSM Evolution), which is the final attempt to push more data through the existing 2G network. It is essentially an improved GPRS and is often referred to as Enhanced GPRS or EGPRS. What EDGE does is to improve the coding used to transmit the binary data. Standard GSM transmission codes bits using a fairly simple frequency shift coding (Gaussian Minimum Shift Keying or GMSK), which essentially allocates two tones to represent a zero and a one. EDGE uses octal Phase Shift Keying (8-PSK), which can cram more bits into the same frequency allocation. In fact, 8-PSK triples the data capacity of each channel, to around 40 to 70 Kbits/s, but it also makes it more sensitive to noise and signal strength and so it also defines nine variations of coding to allow speed to be traded for reliability if the user is on the edge of a cell. Using multiple time slots and with a good signal strength, EDGE can provide nearly 400 Kbits/s.

Both GPRS and EDGE are sometimes referred to as 3G services even though this is misleading. There is one thing that no amount of tinkering can change about 2G services: they all use FDMA and TDMA, both of which are known not to be the best way to share a block of frequencies. The revolution in most wireless technologies such as Wi-Fi has been the use of spread-spectrum techniques. Instead of transmitting on a single frequency, spread-spectrum makes use of all of the available frequencies either by frequency hopping or the even more efficient approach of spreading the signal at low power across the available bandwidth.

When GSM and other 2G services were being designed, spread-spectrum was in its infancy, but one phone system did use a form of it as CDMA, which currently has about 10 percent of the market mostly in the United States. CDMA works by mixing the data bits with a pseudo random bit sequence that makes the resulting signal look like noise; it sounds like background static, which sounds like listening to it on a standard radio. To recover the data, the same pseudo random bit sequence must be subtracted from the received data. The mobile and cell transmitter have to negotiate which of the possible bit sequences to use during their initial setup. After this, different mobiles use different bit sequences, which can transmit using the same frequency allocation at the same time and their data can be recovered.

The European 3G standard—Universal Mobile Telecommunications System UMTS—which is used in 3GSM, makes use of W-CDMA. In this case, the channel used is 5 MHz, which is four times wider than the original CDMA. The 3G data transmission is a packet-based system that is essentially GPRS without the need for time slots. It also uses an adaptive coding system that trades transmission speed for reliability.

Hence, the actual data rate varies according to the strength and quality of the signal. At its best, it allows operators to offer data rates as high as 2.4 Mbits/s. 3GSM is designed to interwork with existing GSM networks, allowing calls to be made on whatever system is available: 2G voice, CSD, HSCSD, GPRS, EDGE, 3G voice, and 3G data. The mobile calls then can be passed between GSM and 3GSM cells and the protocols adjusted to what is available without you knowing anything is happening at all. Different countries have adopted different versions of 3G

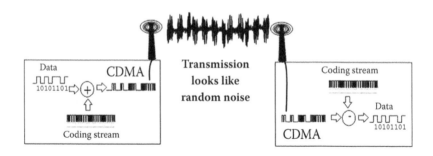

Figure 3.4 CDMA works by mixing the data with a faster stream of coding bits.

CDMA technologies; for example, in the United States, existing CDMA networks have upgraded to CDMA 2000 rather than W-CDMA or 3GSM (see Figure 3.4).

Currently, there are two alternative technologies in use: LTE (Long Term Evolution) and Wi-Max. Wi-Max is just an extension of the Wi-Fi standards to a longer range and a higher power. While Wi-Max is being used to bring the Internet to remote locations or to locations that do not have good ADSL (Asymmetric Digital Subscriber Line) speeds, it is almost certainly not going to be the winner in the 4G stakes. Various companies have produced Wi-Max phones and more are on their way to market but LTE hardware seems much more advanced. If Wi-Max is a development on Wi-Fi, then LTE is a development on GSM/UMTS. It uses advanced signal processing techniques to increase the data rate and reliability. It also works in a different area of the wireless spectrum because its radio is incompatible with 3G signals.

3.2 Machine-to-Machine (M2M) Telemetry

When machines "talk," they do so in a language known as telemetry. M2M started as a utility meter application. M2M uses a device, such as a sensor, mobile, or meter to capture an "event," which is instantly transmitted to a program to take action. M2M applications have been limited to monitoring devices as alert systems; Verizon and Sprint are already doing this in the market sectors of asset tracking, industrial production, transportation management, financial services, utilities monitoring, healthcare alerts, vending maintenance, etc., via mobiles.

The concept of M2M holds great promise in promoting telemetry's use by business, government, and private industry. But most importantly, M2M can help manufacturers, retailers, and other businesses maintain inventory; it make it easier for scientists, doctors, and financial traders to conduct research, manage risk, and take preemptive action. However, the greatest application of M2M is in the area of marketing to mobiles.

What mobile analytics does is elevate the capabilities and functionality of M2M from that of a utility meter to that of a rapidly adaptive and precise mobile marketer. The rules used to issue alerts to divert network traffic for M2M application can now instead be used to discover business rules about mobile behaviors—constructed with decision tree AI (artificial intelligence) software—which can lead to improved sales and revenue.

Mobile analytics is concerned with what mobiles are looking for and where they are located when they do so. As with the original M2M meter applications, it is important for mobile analytics to continuously monitor and model device activities in order to enable conditional adaptive rules to be developed from historical data patterns to generate relevant offers and content in a seamless manner in real-time by triangulating mobiles.

The mobile data boom is already threatening the carrier's once bread-and-butter voice revenues, which have stagnated as people use the Internet for much of their communication, bypassing the use of traditional avenues the cellular companies were built upon. Companies, marketers, and brands were the first to jump on the bandwagon in terms of data, using software and technologies that drive efficiency, revenue, customer loyalty, and profits. Now, the man-in-the-street is doing the same to make their lives a lot easier in an ever-increasingly demanding world.

Emerging academic research suggests that companies that use mobile big data and business analysis to guide decision making are more productive and experience higher returns on equity than competitors that do not, says McKinsey in a recent report (<http://www.mckinsey.com/>). The consumers of today want information, entertainment, and social networking from their mobiles and because of that, telecom companies are transforming themselves in the hope that they can retain their customers for a longer period.

This is also the era of behavioral change and a big data boom, where scalable networks to seamless connectivity is a given. The future is about volume, variety, and velocity, and that is the hallmark of the big data era. Thanks to the booming use of mobiles, the amount of data flowing over cellular and wireless phone networks is growing faster than the providers can keep up with. That is why our cellular companies are in transformation mode. Each is trying to make sure they are in the forefront of the mobile big data growth wave because the growth in the future is in data as voice revenues level off. Their biggest fear for doing that is the loss of their customers to another carrier who can give them what they want.

There are multiple business cases for mobile data offloading, including those for mobile data subscribers and mobile and broadband network operators. The business cases are driven by cost reduction and value creation rather than by new revenue generation. Most mobile data offloading is accomplished via dual-function mobiles that can access a mobile networking service via a 3G/4G wireless interface or independently access the Internet using a Wi-Fi wireless interface. There is no coupling or internetworking between the mobile service and the Wi-Fi-enabled public Internet service. In this configuration, the mobile wireless traffic is backhauled over the mobile operator's owned or leased facilities to the mobile operator's core network, while the Wi-Fi wireless service is typically connected to the operator's broadband Internet access service.

Although the mobile subscriber could manually switch the mobile device between mobile and Wi-Fi service, most mobiles employ a connection manager that can automatically switch to a Wi-Fi network if it detects a known Wi-Fi network. For mobile subscribers, mobile data offloading provides increased coverage and faster downloads, and costs less than mobile service alone. An analysis of 3G smartphone use with Wi-Fi offloading in South Korea sponsored by the U.S. National Science Foundation (NSF.gov) monitored the usage of 100 iPhone subscribers for 24 hours per day for 2.5 weeks. This research provided data on the timing and mix of service availability for urban subscribers, including access to home Wi-Fi, 3G mobile, WLAN at work, and public Wi-Fi in retail areas and transportation facilities. The study found that about 65 percent of the total traffic was offloaded using on-the-spot Wi-Fi access. It found, further, that an additional 29 percent of the total data load would be offloaded if offload delays of at least an hour were permitted. From the subscriber's perspective, mobile offloading will go a long way in keeping 3G data usage below mobile operators' monthly data usage caps.

The study measured an inverse relationship between 3G and Wi-Fi coverage: 3G has better availability outdoors while Wi-Fi has better indoor availability. Overall coverage therefore is enhanced by mobile data offloading. Average end-to-end data rates were measured as 1.26 Mbits/s during the day and 2.76 Mbits/s during the night. The data rate is higher during the night because smartphones have access to subscribers' home Wi-Fi systems. The mobile operator's mobile data offloading business case is built upon cost avoidance. Mobile data revenue growth, although very strong, is not keeping pace with traffic growth.

For example, AT&T (NYSE:T) reports that in its most recent quarter (ATT. com 2012, 3rd quarter), mobile data revenue increased by 18 percent from the same quarter in the previous year. Over the same period, AT&T reported a 54 percent increase in multimedia messages. This would imply traffic growth in excess of 100 percent. Although mobile operators have aggressive investment programs to expand the capacity of the Radio Access Networks (RAN) by upgrading and deploying additional cell sites and by adding capacity to mobile backhaul facilities, it seems unlikely that a business case can be made to completely close the gap between the cost of facilities needed to support traffic growth and expected revenue growth by making further investments in owned or leased network capacity.

Mobile data offload is an inexpensive way to provide the extra capacity needed to support the expected traffic growth. The mobile operator's only cost is that of subsidizing the dual-function (3G/4G and Wi-Fi) devices sold to subscribers. There also is a potential opportunity cost to the mobile operator in that revenue from data traffic in excess of the monthly capacity limit is forgone by offloading. However, subscribers are highly resistant to paying excess usage charges and are more likely to cancel service or limit its use. Either behavior is harmful to the mobile operator's business case, in that even if the service subscription is not canceled, the service's perceived value to the subscriber is impaired if usage is curtailed. Mobile offloading, therefore, increases the subscriber's perceived value of the mobile data service while improving the mobile operator's profitability.

The business case for the broadband service provider used to backhaul the Wi-Fi service is more nebulous. Most mobile data offloading occurs in subscribers' homes. Residential mobile data offloading adds to total broadband traffic without making any revenue contribution to the broadband operator. However, mobile data offloading does increase the value (utility) of broadband service to the mobile data subscriber. Mobile data offloading, consequently, makes residential broadband more attractive to consumers and therefore increases service stickiness and improves broadband subscribers' willingness to pay.

Mobile data broadband is a rare telecom service in that it delivers increased value to mobile data subscribers, mobile operators, and broadband service providers. Additional spectrum must be urgently identified and allocated to support the growth in mobile data. As non-voice traffic grows explosively over the next decade, the UMTS Forum (UMTSForum.net), a mobile industry association, warns that operators face increasing challenges to add capacity and coverage to their existing networks. As the efficiency of broadband cellular technologies rapidly approaches theoretical limits, the UMTS Forum (UMTSForum.net) says the gap between demand for media-rich mobile services and available network capacity will increase sharply in the coming decade.

In its most recent projections, the UMTS Forum has calculated that mobile data traffic will grow by a factor of 33 percent from 2010 to 2020. This growth is even more pronounced in Western Europe, where the same study forecasts that data traffic will leap by a factor of 67 percent in the same ten-year period. Investment in new technologies and network density alone cannot address this dramatic increase in demand, the Forum warns.

To sustainably deliver the full socioeconomic promise of mobile broadband, it is clear that advances in technology and investment must be complemented by timely availability of harmonized radio spectrum to support new services and more users. The UMTS wants to put pressure on the International Telecommunication Union (ITU) to exercise its global leadership position by putting IMT (International Mobile Telecommunications) spectrum high on the agenda at the next World Radiocommunication Conference in 2015.

3.3 Mobile Data Metrics

The mobile miner needs to map a strategy and set clear measurable objectives. The mobile marketer must be fully aware of the available software, networks, and solution providers in order to execute their tasks and ad campaigns. The mobile analytics metrics need to implement a framework for leveraging these data streams of mobile behaviors as the building blocks for the following goals: (1) increase sales, revenues, and profits via improved customer service and content relevancy, (2) measure cross- and up-selling ratios, and (3) measure the improvement of the consumer experience and loyalty.

Every enterprise has streams of transactional and behavioral mobile device data flowing to it 24/7 but few are able to triangulate them simultaneously as events take place, thus enabling them to make relevant offers to their existing and new customers; at the moment, they interact with them via their apps and mobile websites. With every consumer event, mobiles are communicating their needs and desires to brands, companies, and marketers. Mobile analytics can leverage these digital device events, most of which start at mobile sites; the following is a short list of critical items to consider.

- *Log analyzers*: The simplest kind of mobile analytics software are analyzers of Web log server files, which can yield important indicators such as by whom, when, and how a mobile Web server is visited. Some of the common tracking features by these analyzers include Short Message Service (SMS), which are text messaging service components; activity in mobile websites, the tracking of customized mobile apps; Quick Response (QR) Codes; and Multimedia Messaging Service (MMS*)*, which is a standard way to send messages that include multimedia content to and from mobiles.

  An industry leader in this area of mobile site log analysis is Webtrends® (Figure 3.5), which is experiencing explosive growth in the measurement and reporting of mobile activities due to the intense focus of such client brands as Coca-Cola®, Microsoft, Monster®, Orbitz®, RIM®, Motorola®, and Mutual Mobile®. Unfortunately, IP geo-location of mobiles can expose different IP

Figure 3.5   Mobile metrics from Webtrends®.

addresses to servers within time spans of a few minutes, thereby rendering IP-based user tracking by these analyzers impossible; however, other options exist because of the triangulation needs of mobiles.

■ *Social mobile marketing*: This is the orchestrated strategy to communicate and coordinate promotional elements—and also the center of efforts to create content that attracts attention and encourages users to share product and service content with other mobiles. A social media message spreads from user to user and resonates because it is coming from a trusted third-party source, as opposed to coming from a brand or company itself. Social mobile media is a platform that is a relatively inexpensive methodology to implement with strategically higher value.

■ *Mobile search engine optimization (MSEO):* This is the process of improving the visibility of a mobile site in Google, Yahoo!, or Bing. MSEO techniques and strategies involve the optimization of the popularity and prominence of a mobile site by the use of keywords for the autonomous indexing of search engine spiders. For example, search engine optimization is a method of getting a mobile site to rank higher in Google.

■ *Model Web device behaviors*: This is the study and modeling of when, why, how, and where people do or do not buy a product or service. It blends elements from psychology, sociology, social anthropology, and economics; it is an effort to understand the buyer decision-making process, both individually and in clusters of mobiles. It studies the characteristics of mobiles behavior variables in an attempt to understand users' desires and preferences. It also tries to assess influences on the consumer from groups such as friends, other mobiles, and social media.

■ *Mobile cookies, beacons, and apps*: The strategic use of these Internet and wireless mechanisms are vitally critical; they represent the components for ensuring the capturing and modeling of mobile device behaviors for predicting consumer preferences, desires, and needs. For specifics on how to develop mobile Web cookies, see http://www.w3.org/TR/mobile-bp/.

■ *Digital fingerprinting*: They are coded string of binary digits, generated by a mathematical algorithm, that uniquely identifies a data file. Digital fingerprinting is used in detecting the tampering of electronically transmitted messages and, more importantly, can be used to track and identify mobiles for advertising purposes. Digital fingerprinting, similar to an analog fingerprint of a person, cannot be reconstructed because no two files can differ even by a single character as they will have a completely different digital fingerprint, which for the marketer means a more precise level of behavioral targeting.

One free and easy way to obtain mobile metrics is via the use of Google Analytics, which makes it possible to view the ads performance of mobiles devices. Google makes available JavaScript-based GoogleAnalyticsTrackingCode that empowers sites to track activity on standard webpages; it provides a Web service

Mobile Analytics

Understand how mobile impacts your site and how it drives business to you. Measure mobile websites, mobile apps and visits from web-enabled mobile devices, including both high-end and basic phones. Measure ads that lead people to use your app and find out whether they prefer ads on their desktop or mobile. You'll then be setup to create targeted and efficient marketing campaigns that reach your visitors wherever they are.

Figure 3.6   Mobile Analytics from Google.

and a pair of SDKs to help track user interaction at mobile sites and apps usage accessed from mobiles.

The Google Analytics for Mobile Apps SDKs make it easy for developers to track the performance of their mobile apps. The SDKs can be downloaded and used with the Apple and Android application platforms. The SDKs enable developers to identify the places in an app where it is likely to trigger a visit, page view, or an event—such as a purchase. Google Analytics (Figure 3.6) allows brands and companies to notice the number of visits that they receive as a result of mobile ads, which can help them gain even more insight into their consumers and how they access their mobile site; they can even see which mobiles their visitors use and optimize their sites for those mobiles.

3.4 Mixing Personal and Business Mobile Data

Mobiles now make up almost half of all device sales, and many millions of people use them for personal communications as well as work-related tasks. Yet, few people want to carry multiple mobiles at all times just to separate the workday from their home life. Here are several ways a user can access their business and personal accounts without having to carry two mobiles.

One method of making a personal mobile safe for work data is through software installed onto a user's personal phone by the employer. Mobile Device Management (MDM) software such as Microsoft System Center™ (<http://www.microsoft.com/en-us/server-cloud/system-center/default.aspx>) can control settings on their mobile, making sure their security setup is safe for work data. Another option is virtual machine software such as VMware® MVP™ (<http://www.vmware.com/>), demonstrated at CES on an Android-powered LG® phone, which allows a mobile to maintain a separate business environment with its own apps and settings, like a mobile within a mobile, making a clear distinction between work and personal data; these solutions are still very new.

Similar to using a virtual machine, there are many apps that allow a mobile to access a Mac or PC, letting the user see and control the programs running on the faraway computer. Examples include TeamViewer™ (<http://www.teamviewer.com/>) on iOS and Android, or LogMeIn® (<https://secure.logmeinrescue.com/>) on Apple and LogMeIn Ignition (<https://secure.logmein.com/products/ignition/>) on Android. Such remote desktop apps provide another way to have a clear line between work and personal data, but this method relies heavily on having a reliable data connection. Without that, the user will not be able to see his or her work computer or access any of its programs or data.

For some purposes, there are ways to access the data from one type of account via another account. For example, using forwarding and filters, it is possible to forward messages from a work email account automatically to a specific folder or label in a personal account. This is also common on calendars, where users could give their Google Apps for Business™ Calendar work account full proxy access to allowing them to see and manage it through their personal Google Calendar™. They can even use Google Voice™ to forward calls from one of their numbers to the other. Some of these options are available through other systems as well, including Microsoft Exchange with Outlook®.

LogMeIn Ignition is like having multiple accounts, similar to using multiple login profiles on a computer. Although Android mobiles do not provide multiple profiles, they do allow syncing with multiple Google accounts. The Google Calendar app uses this to display calendars from multiple accounts, all commingled on the same screen. The Google Gmail™ app does something similar but keeps the data from multiple accounts separate, allowing the user to switch among them. They can even have different notification sounds for each account, so the user knows by the tone when work or personal email arrives. Windows Phone 7.5 just added some limited syncing options as well, allowing access to twenty-five Google calendars on a mobile, and a "Send Mail As" option that can include the work email address.

For those instances where one app will not work for both work and personal purposes, then two apps will have to do. On any mobile, this is common for email, where the user might use the Gmail or Yahoo! mobile apps for personal mail, and the Email app, which can use Microsoft Exchange ActiveSync™, for access to the company's Exchange server. If the user wants to keep his or her Web browser bookmarks, cookies, and cache separate, he or she can use the default mobile browser for personal while installing another browser, such as Dolphin® (<http://dolphin-browser.com/>), on Android and Apple mobiles for work.

In general, commingling personal and business mobile data is not a good idea, and many companies have policies against accessing their work mail through a personal email account. In a business where data security is vital, MDM or virtual machine software is best. If the user always has a strong data connection from his or her mobile, a remote desktop app will also keep sensitive data safe. Short of that, most people are likely to have a mix of apps that support multiple accounts, such as Gmail on Android for email, while using multiple apps for other activities where one app cannot do both.

3.5 GPS and Wi-Fi Triangulation

Two of the general criticisms of GPS technology are that it does not do very well in dense urban areas and it is not accurate enough for mobile targeting. SkyhookWireless™ (<http://skyhookwireless.com/>) took note of these drawbacks several years ago and began to develop new targeted Wi-Fi–based technology that can go to more places and be, by far, more accurate than GPS.

Skyhook began to sniff out and map Wi-Fi networks, measuring their signal strength and building the results to their proprietary database. Skyhook continues to conduct periodic sweeps to catalog Wi-Fi hotspots; their scans do not actually connect to or use any of these hotspots—they only scan them to measure relevant factors used for their database lookup and updates.

Wi-Fi technology is strong and accurate enough to navigate inside buildings, so much so that Apple used Skyhook's commercial offerings in the early versions of its mobiles. The Skyhook technology allows for tracking and targeting of mobiles within ten meters. Apple, in later mobile device releases, created and used its own proprietary location targeting database. However, Skyhook's publicly available Core Engine™ software development kit (SDK) allows app developers and mobile marketers to quickly and easily integrate Skyhook's location-based system on the platform of their choice. The Skyhook Core Engine supports Android (Google), Linux, Mac OS X (Apple), Symbian, Windows Mobile, and Windows 7. Skyhook developed the Core Engine, a software-only location system based on a combination of Wi-Fi positioning, GPS, and cell tower triangulation.

The Skyhook Core Engine is fast and accurate, reliable and flexible, and supports multiple devices and mobile applications. Skyhook patented technology enables hundreds of millions of mobile applications to quickly determine a device's location within meters. A mobile device with the Skyhook Core Engine collects raw data from each of these multiple location sources. The Skyhook client then sends this data to the Skyhook Location Server and a single location estimate is returned. The client is optimized so that it communicates with the Skyhook Location Server only when the location cannot be determined locally. This behavior minimizes the user's data cost while maximizing the battery life of mobiles. Skyhook maintains the accuracy of this database through an ongoing and continuous process of data monitoring, analysis, and collection. The three main components include

1. *Baseline data collection and establishing coverage areas*: Data collection begins with identifying target geographic areas using population analysis. Skyhook territory planners build coverage schedules starting with population centers

and then moving into residential and suburban areas. Skyhook deploys a fleet of data collection vehicles to conduct a comprehensive access point survey within the target coverage areas in search of Wi-Fi hot spots.

2. *Automated self-healing network*: As more users reference location for mobiles, the Skyhook database is automatically updated and refreshed.

3. *Periodic rescan*: Depending on the aging of the survey data and the density of user-generated updates, Skyhook periodically will rescan entire coverage areas to recalibrate the reference network, ensuring that performance remains consistent over time. Every territory added to Skyhook coverage is continuously monitored to assess the quality of the reference network and determine whether a rescan is required.

Wi-Fi positioning performs best where GPS is weakest—that is, in urban areas and indoors. GPS provides highly accurate location results in "open-sky" environments, like rural areas and on highways. But in urban areas and indoors, tall buildings and ceilings block GPS' view of satellites, resulting in serious performance deficiencies in time to fix on to mobiles. GPS or A-GPS alone cannot provide fast and accurate location results in these environments. Cell tower triangulation provides generalized location results with only 200- to 1,000-meter accuracy; it serves as a coverage fallback when neither GPS nor Wi-Fi is available.

Skyhook, Google, and Apple maintain worldwide databases of cell tower locations, which increases their coverage areas and helps improve GPS satellite acquisition time. Up until 2010, Apple relied on databases maintained by both Google and Skyhook to provide location-based services for their mobiles running their iOS versions 1.1.3 to 3.1. However, beginning with iOS version of 3.2 for their devices, they switched to their own Apple databases to provide location-based services and for diagnostic purposes. These databases must be updated continuously to account for changes in the physical and digital landscapes.

Apple maintains that their devices are not being targeted for their location. Rather, they are maintaining a database of Wi-Fi hotspots and cell towers to assist their mobiles in rapidly and accurately calculating their location when requested. Apple states that it is not storing device location data but rather a subset (cache) or the crowd-sourced Wi-Fi hotspot and cell tower databases that are downloaded from Apple to their devices. The Android operating system offers more granular controls and permissions for how their mobiles behave than does Apple; BlackBerry devices also offer options for users to block certain types of location-based services entirely. The Apple End User Software License Agreements (SLAs) for their iPhone device products states the following:

> "Apple and its partners and licensees may provide certain services through your iPhone that rely upon location information. To provide these services, where available, Apple and its partners and licensees may transmit, collect, maintain, process and use your location data,

including the real-time geographic location of your iPhone, and location search queries. The location data collected by Apple is collected in a form that does not personally identify you and may be used by Apple and its partners and licensees to provide location-based products and services. By using any location-based services on your iPhone, you agree and consent to Apple's and its partners' and licensees' transmission, collection, maintenance, processing and use of your location data to provide such products and services. You may withdraw this consent at any time by not using the location-based features or by turning off the Location Service setting on your iPhone. Not using these location features will not impact the non location-based functionality of your iPhone. When using third party applications or services on the iPhone that use or provide location data, you are subject to and should review such third party's terms and privacy policy on use of location data by such third party applications or services."

Similar provisions regarding location-based information appeared on all other Apple mobiles, including their iPhone 4®, iPad®, iPod touch®, Mac OS X®, and Safari 5® SLAs. Apple revised their SLA in 2007 to update customers about the necessary exchange of information between its servers and their devices. In 2008, the SLA was updated again to include the use of "pixel tags" at its iTunes® site. Pixel tags, also known as "beacons" or "bugs," are tiny graphic images used to determine what parts of a site a visitor navigates to; they are silent tracking tags used to measure the activities and behaviors of mobile users.

That same year, Apple began to provide location-based services enabling apps to track and allow mobiles to perform a wide variety of tasks, such as getting directions to a particular physical address from their current location, as well as locating their friends' devices and letting them know where they are, or identifying nearby restaurants or stores and other social media functions.

In 2010, Apple updated its SLA yet again, to incorporate its location-based services to all of its mobiles. It updated the policy to incorporate the provision regarding new Apple services, such as their new iAd Network. Most of the revisions dealt with how the company can use mobile cookies to preserve and protect the information of children and international customers.

To provide location-based services, Apple, Google, and Skyhook must be able to determine quickly and precisely where a device is located. To do this, all three must maintain secured databases containing information regarding known locations of cell towers and Wi-Fi access points. They each store this information in their proprietary databases accessible only by them and yet do not reveal personal information about individuals.

Information about nearby cell towers and Wi-Fi access points is collected and sent to Apple, Google, and Skyhook with the GPS coordinates of mobiles; and if available when an individual requests current location information, it automatically, in some cases, updates and maintains databases with known location information. In both cases, devices collect the anonymous information about cell towers and Wi-Fi access points.

All three companies collect information about nearby cell towers, such as the location of the tower(s), Cell IDs, and data about the strength of the signal transmitted from the towers. A Cell ID refers to the unique number assigned by a cellular provider to a defined geographic area covered by a cell tower in a mobile network. Cell IDs do not provide any personal information about mobile device users located in the cell.

Apple, Google, and Skyhook also collect information about nearby Wi-Fi access points, such as their location and Media Access Control (MAC) addresses, and data about their strength and speed of the signal transmitted by the access point(s). A MAC address is a unique number assigned by a manufacturer to a network adapter or network interface card (NIC) of a mobile device. The address provides the means by which a device is able to connect to the Internet. MAC addresses do not provide any personal information about the owner of the NIC. None of the three companies collects the user-assigned name of the Wi-Fi access point, also known as the service set identifier (SSID) or data being transmitted over the Wi-Fi network, also known as the "payload data."

Because Apple began to provide location-based services in 2008, it handles customer requests for current location differently. The device's GPS coordinates, when available, are encrypted and transmitted over a secured Wi-Fi Internet connection to Apple; for devices running their iOS version 3.2 or iOS4, Apple will then retrieve known locations for nearby cell towers and Wi-Fi access points from its own proprietary database and transmit the information back to the mobile.

However, for requests transmitted from devices running *prior* versions of the iOS, Apple transmits—anonymously—the Cell Tower Information to Google and Wi-Fi Access Point Information to Skyhook. These providers return to Apple known locations of nearby cell towers and Wi-Fi access points, which then transmits the coordinates back to the mobile. The device uses the information, along with GPS coordinates, if available, to determine its actual location.

Apple automatically collects this information only if the device's location-based service capabilities are toggled to "On" and the customer uses an application requiring location-based information. If both conditions are met, the mobile device intermittently and anonymously collects cell tower and Wi-Fi access point information from the cell towers and Wi-Fi access points that it can "see," along with the device's GPS coordinates, if available. This information is batched and then encrypted and transmitted to Apple over a Wi-Fi Internet connection every 12 hours, or later if the device does not have Wi-Fi access at the time.

Finally, the Federal Communications Commission (FCC) will require that mobile device providers and Voice over the Internet Protocol (VoIP) providers ensure that their products meet stricter standards for location accuracy than currently apply to mobiles with GPS capability. The FCC intends for these standards to be implemented by a date to be determined beginning after 2019. Providers will be able choose a device-based system with an imbedded GPS-type chip, a network-based system, or a hybrid of the two. The new regulation will allow almost universal pinpoint location of 911 callers by emergency responders; the FCC estimates that by 2018, 75 percent of all mobiles will be GPS capable.

Navigating large stores such as IKEA® and shopping centers such as the cavernous Mall of America® can be a daunting task. Up until now, the powerful mobiles have been useless for indoor navigation due to poor GPS signal quality while beneath a thick slab of concrete. Indoor location navigation technology is on the way from companies such as Wifarer™ (<http://www.wifarer.com/>); in addition, Google and Nokia are both working to remedy the situation with the development of inside positioning systems (IPSs). The wireless technology behind IPS is nothing new, as it makes use of Wi-Fi and Bluetooth in conjunction with cell towers to triangulate a user's position, rather than satellites in orbit around the Earth. Using those technologies with accurate cartography of inside locations enables companies to provide location services that help consumers get to where they want to go.

The most widespread use of IPS comes in the form of Google Maps™. Mountain View Maps team™ recently turned its attention to large structures such as the aforementioned Mall of America and IKEA to test the reliability of its IPS process. As the mobile enters one of those locations from street side, the Maps app on Android starts polling wireless routers that are placed around the area to compare coordinates with cell tower readings. These routers are strategically placed around the entire building, handing users off to each other to keep locations synced, with an accuracy of up to five meters—which is not bad at all.

Nokia, on the other hand, is claiming they are able to pinpoint a mobile within a foot using Bluetooth 4.0 to communicate with devices. As yet unavailable to consumers, this system relies on much more hardware than does the Google method. Bluetooth does not have great range, so to be able to provide location information relative to position, there would have to be quite a few Bluetooth transmitters to ensure coverage. Whether or not retailers are going to be willing to bear the brunt of the extra cost of hardware in exchange for accurate locations remains to be seen.

Another company, Sensewhere® (<http://www.sensewhere.com/>), is taking an altogether different route to tackle this issue. Sensewhere takes the stance that a Wi-Fi signal is too undependable to rely on for location, so it has developed an app that measures the changes in a mobile's radio environment as the user walks around. As the spectrum makeup changes, the app can relate that information to a location on a map.

There are hurdles to all the IPS technologies listed above. As users know from having Wi-Fi in their home, signal quality can degrade with the tiniest change in their location, or the position of furniture or appliances. Imagine trying to administer a router in a world-class shopping mall that has tons of people and changing storefronts. It could be a nightmare to ensure that everything stays up and running, but all these companies are working on the problem of accurately locating mobiles.

3.6 Backing Up Mobile Data

One of the tasks for data mining mobile devices is ensuring that mobile data is adequately available for behavioral analysis. Mobiles have become an integral part of consumers and businesses, and have successfully replaced several essential attributes and mediums, which users have traditionally relied upon. Mobiles are often stuffed with all the data, be it contacts, calendars, events, reminders, photos, messages and more. For example, for a business user who stores important client information, this is vitally important.

However, what if that user's mobile is stolen or lost. Getting a replacement could just be a matter of time, but the data is something that cannot be replaced. Here is when taking a backup of all that data becomes essential, especially when the user relies on contact details and messages of utmost importance, let alone photos, business intelligence, and videos. The only solution to this is appropriate and timely backup of all mobile data in a secure place, which can then be easily restored to their newer devices.

Email services come pretty handy while taking mobile backups. Most users access their email every day, so backing up mobile data using the various services is simple. Google and Yahoo! accounts will let developers sync their contacts easily, be it Android, Apple, BlackBerry, or Windows. Google Mobile sync lets the user or developer back up all data from all of these mobiles. Companies such as BlackBerry go a step further with backup provided through its Enterprise™ server and also by offering the BlackBerry Protect™, which is a free application that automatically backs up contacts, text messages, calendar, and bookmarks via mobiles, which probably is why traditionally BlackBerry is preferred by the corporate sector. In addition to email services and BlackBerry in-house backup options, users can also opt for some online syncing services that help them back up their mobile data. Although still debatable, cloud services could be a great way to back up mobile data.

Mobile giants have introduced their own cloud services; for example, Apple has iCloud™ for all its iDevices, and there is Skydrive™ from Microsoft and Google cloud service. Rseven is an online solution that saves mobile data, like call logs, email, contacts, messages, video, audio, images, and calendar online on its secure website. It can even record calls and play back voicemail on the website. It easily backs up all this mobile data when users switch to another mobile. In addition to Dropbox® for Apple, Android, and Blackberry mobiles, SugarSync™ also supports these mobile platforms and so does iDrive, which offers great options of securing mobile data. With a limited storage space initially, mobile developers, users, and analysts can buy additional space.

Saving data on the SD (secure digital) card comes across as one of the simplest and most cost-effective ways of taking a timely backup of mobile data. All contacts can be saved on the SD card and then imported into a different mobile. Google popularly uses the .csv format to import and export contacts on their Android mobiles as well as their Gmail service. One can effortlessly back up data on the memory card via his or her "Contact" settings. There are also SIM (subscriber identity module) card readers available that allow the user to back up the SIM data. However, SIM memory is limited and the user and developer may have to delete some messages frequently.

3.7 Real-Time Demographics

One of the goals of data mining mobile devices is the desire to offer, with precision, relevant content and ads about specific products and services to mobiles. One possible strategy is to infer demographic information, such as gender, age, or marital status, about owners of mobiles who use customized apps or while visiting a mobile website. Mobile device behaviors are usually anonymous; they nonetheless provide a certain amount of usage information, such as search terms entered by the mobile user, which reveal the device owner's interests.

When users download apps to their devices, registration is usually required prior to activation; so, for example, the Pandora app will ask for the user's gender and age, which when coupled with location-based information, can be used to develop real-time demographic usage profiles. For strategic mobile mining, once collected, some of this big data usage information can be used to construct predictive models to anticipate what products, services, or content to offer mobile users of apps or at mobile sites.

Table 3.1 contains some of the anonymous real-time demographics that can be used for predictive modeling via data mining mobile devices based on information collected from consumers.

There are also commercial real-time demographic networks, such as Acxiom Relevance-X®, which can provide demographics and lifestyle consumer information in real-time for association to mobiles. The Acxiom demographics can be used to enhance mobile analytics performance; enterprises and mobile marketers can

Table 3.1   Anonymous Real-Time Demographic Information

Variable	Value
Gender:	male, female
Age–18	true, false
Age 18–34	true, false
Age 35–54	true, false
Age 55+	true, false
Marital status	single, married

subscribe to it in order to improve their relevance and sales. A related product is Acxiom Relevance-X® Social, which offers consumer data intelligence for social mobile marketing.

Acxiom Relevance-X Social helps marketers see the social networks of mobiles and how many friends or contacts they may have; it is available for individual ad campaigns or as an ongoing service. With Acxiom Relevance-X Social data, mobile marketers, developers, and analysts can

1. Establish up-to-date social intelligence on their visitors and their mobiles.
2. Invite device influencers in an engaging way to drive purchase behavior.
3. Plan media where customers and their mobiles are socially active.
4. Interact with socially active brand advocates and their mobiles.
5. Test new products or services on multi-channels to all mobiles.
6. Develop loyalty programs to reward mobile device segments.
7. Create mobile campaigns that solicit user-generated content.
8. Identify mobiles that demonstrate brand enthusiasm.

Mobile data traffic from devices over the Internet is skyrocketing; Cisco Systems® has come up with a forecast that is downright staggering: mobile data globally is expected to grow 18-fold by 2016. Cisco (Cisco.com) said in the forecast update to its VisualNetworking Index that traffic alone will grow to 50 times more than it is today; by 2016, there will be more mobiles (10 billion) than the number of people on Earth, estimated at 7.3 billion. The sharp increase in mobile data is primarily because of the increase expected in new mobiles.

Cisco said the amount of mobile data traffic was about 0.6 EB (exabytes) per month in 2011, but that figure will reach 10.8 EB per month in 2016. That is an increase of 78 percent per year. By 2016, the annual amount of mobile data will be 130 EB, equal to 33 billion DVDs, or 4.3 quadrillion music files, or 813 quadrillion text messages. If that is not enough, there will be so much more mobile data traffic added between 2015 and 2016 that the amount added in that year alone will be three times the estimated size of the entire mobile Internet in 2012. The growth of mobile data, coupled with the increase in real-time demographics, will make the data mining of mobile devices an important component to mobile analytics and marketing.

3.8 Hyperbolic Positioning

Geo-location via triangulation can assist mobile developers and marketers in knowing where revenue-producing devices are coming from, which may impact how and what ads and offers to make. Geo-location can provide meaningful location information—such as a street address or zip code—rather than just a set of geographic coordinates. Specifically, this involves the use of advanced radio-frequency (RF) location systems utilizing, for example, time difference of arrival (TDOA), also known as *hyperbolic positioning*, which is the process of locating an object by accurately computing the TDOA of a signal emitted from that stationary or mobile device to three or more receivers; this triangulation offers great specificity of mobile device location to analysts, developers, and marketers.

TDOA systems often utilize mapping displays or other graphical information systems. In addition, mobile device geo-location can be performed by associating a geographic location with its IP address—or a Media Access Control (MAC) address, which is assigned by device manufacturers of NICs that are stored in device hardware, the card's read-only memory, or some other firmware mechanism. If assigned by the manufacturer, a MAC address usually encodes the manufacturer's registered identification number and may be referred to as the *burned-in-address*.

It may be a MAC address or radio-frequency identification (RFID), which is a technology that uses communication through the use of radio waves to exchange data between a reader and an electronic tag attached to a device for the purpose of identification and tracking it. RFID hardware is an embedded device number, embedded into such software as a universally unique identifier (UUID), exchangeable image file format (Exif) as those found in digital cameras, or a Wi-Fi connection location or the mobile device's GPS coordinates.

DigitalEnvoy® and their Digital Element™ service delivers *IP intelligence* and *geo-targeting,* which is used by most of the world's largest ad networks and publishers for noninvasive IP intelligence for targeted advertising, content localization, geographic rights management, behavioral analytics, and local search. Geo-targeting is the practice of customizing an advertisement for a product or service to a specific market based on the geographic location of potential buyers.

Mobiles connected via Wi-Fi are all targetable down to a city level, using Digital Envoy IP targeting technology. Mobiles connected via mobile networks are also targetable but can be problematic to acquire with any precision due to proxy issues and other network limitations. Digital Envoy collects only a user's IP address in its privacy-invasive techniques; alien probes, cookies, and intrusive scripts are never used in their data collection methodology.

For geo-targeting mobiles, mobile marketers and app developers can insert an SDK from a mobile ad network, such as Greystripe Inc.; this is a common practice among app makers, who use these ready-made SDKs to place ads and generate revenue. Greystripe Inc. uses geo-location to locate a mobile device by identifying its Internet address; that is common among websites, but less so on mobiles. Most apps use GPS satellites or maps of triangulated Wi-Fi hot spots to locate users and their devices.

Finally, the chip-making company CSR is adding new features for indoor tracking and navigating to its line of SiRFusion™ location technology for multiple mobiles. The company's SiRFusion platform and its SiRFstarV™ mobile chip architecture amount to the latest navigation technology that customers could use in mobiles to track a person's location as he or she is walking through a large building such as the New York-New York Casino in Las Vegas.

This is important because hyperbolic positioning clearly indicates that location has quickly gained traction with consumers as an essential contextual component for many of the applications they use and depend upon with their mobiles. The motion-tracking system for indoors puts together as good a fix on a person's location as it can with three technologies: (1) navigation data from the GPS and other satellite info services, (2) Wi-Fi radio network triangulation, and (3) motion-sensing devices inside the mobiles, such as gyroscopes and compasses.

The sensors inside a mobile device can offer location clues as well. A compass can indicate in what direction the user is moving; motion sensors such as gyroscopes can tell whether or not the device is moving. The CSR platform is a cloud computing solution, mashing up a bunch of data sources and computing them together to get some really useful information. Because the data can be computed in real-time, it can show whether a user is walking inside a large casino or driving inside its parking lot.

3.9 Privacy Concern of Mobile Data

Living in a world of social networking and mobiles means trading away some personal information; but assessing the price of admission to join the super-networked, digital class is not so simple; even experts on the issue admit that they do not have a full picture of the way personal information is collected and used on the Internet. But here are some basic guidelines to keep in mind. Social networks such as Facebook and Google+™ require, at a minimum, that the user provide them with his or her name, gender, and date of birth. Many people provide additional profile information, and the act of using the services—writing comments or uploading photos or "friending" people—creates additional information about users. Most of that information can be kept hidden from the public if users choose, although the companies themselves have access to it.

If the users have Facebook credentials to log-on to other Web sites, or if they use Facebook apps, they might be granting access to parts of their profiles that would otherwise be hidden. Quora (<http://www.quora.com/>), for example, a popular online Q&A site, requires that Facebook users provide it access to their photos, their "Likes," and information that their friends share with them. TripAdvisor® (<http://www.tripadvisor.com/>), by contrast, requires only access to "basic information," including gender and lists of friends. Social media apps on mobiles that have access to personal phone call information and physical location put even more information at play.

On Apple mobiles, apps must get user permission to access GPS location coordinates, a procedure that will now be applied to address book access as well after companies, including Twitter, were found to be downloading address book information from mobiles. Beyond those two types of data, Apple locks away personal data stored in other apps, such as notepads and calendars. The Google Android operating system allows third-party apps to tap into a bonanza of personal data, although only if they get permission. In order to download an app from the Android Market, users must click "OK" on a pop-up list that catalogs the specific types of information that each particular app has access to.

With both mobile and Facebook apps, often the choice is to provide access to personal information or not use the app at all.

Personal information is the basic currency of an Internet and mobile economy built around marketing and advertising. Hundreds of companies collect personal information about Web users, slice it up, combine it with other information, and then resell it. Facebook does not provide personal information to outside marketers, but other websites, including sites that access Facebook profile data, may have different policies. The data that third-parties collect is used mainly by advertisers, but there are concerns that these profiles could be used by insurance companies or banks to help them make decisions about who to do business with.

In the United States, federal law requires websites that know they are being visited by children under 13 to post a privacy policy, get parental approval before collecting personal information on children, and allow parents to bar the spread of that information or demand its deletion. The site operators are not allowed to require more information from the children than is "reasonably necessary" for participating in its activities.

For those who are 13 or older, the United States has no overarching restrictions. Traditional and mobile sites are free to collect personal information, including real names and addresses, credit card numbers, Internet addresses, the type of software installed, and even what other sites people have visited. Sites can keep the information indefinitely and share most of what they get with just about anyone.

Sites are not required to have privacy policies. Companies have most often been tripped up by saying things in their privacy policies—such as promising that data is kept secure—and then not living up to them. That can get them in trouble under the federal laws against unfair and deceptive practices. Sites that accept payment card information must follow industry standards for encrypting and protecting that data. Medical records and some financial information, such as that compiled by rating agencies, are subject to stricter rules.

European privacy laws are more stringent, and the European Union is moving to establish a universal right to have personal data removed from a company's database—informally known as the "right to be forgotten." That approach is fervently opposed by companies dependent on Internet advertising. The year 2011 saw a flurry of activity on Capitol Hill as U.S. lawmakers introduced a handful of do-not-track bills, with even the Obama White House calling for a "privacy bill of rights." Still, with half a dozen privacy laws meandering through Congress, most observers expect it will be a long time before any are passed—and not before they are significantly watered down in the legislative process.

3.10 Deep Packet Inspection (DPI)

Cookies and beacons—GPS and Wi-Fi triangulation—make it possible for mobile analytics to deliver highly relevant content and ads to mobiles based on their behaviors. All of these tracking mechanisms can be used to enhance mobile analytics, including deep packet inspection (DPI) technology. DPI examines the transmitted data in seven layers of packets rather than just their headers. DPI allows for the extraction of mobile device behaviors at a granular level—such as what sites a device has visited, information about page content, durations of visits, search engine used and their mobile ID—directly from streaming packets.

The use of DPI for triangulating mobiles was made possible by a U.S. Government wiretapping law known as the Communication Assistance for Law Enforcement Act (CALEA) passed in 1994. CALEA was originally intended to provide the ability of the FBI to conduct surveillance of VoIP communications. The core technology allowing for the decomposition of Web and mobile communications for CALEA is DPI, which can be used for the construction of very sophisticated models of mobile device behaviors.

DPI devices and service providers have the ability to look at multiple layers inside packets, such as Layer 2 through Layer 7 of the Open Systems Interconnection (OSI) model data standard, which is a description for layered communications and computer network protocol design (Figure 3.7). For example, DPI can tell if the packet is from an Apple, Android, or Skype™ device; all of this vital information can be used for segmentation via mobile analytics. DPI of mobiles uses the GPRS Tunneling Protocol (GTP), which is a group of IP-based communications protocols used to carry GPRS that can be used to track a mobile device location with such metrics as time/duration and over fifty other metadata Mobile Traffic Optimization (MTO) features that can be used for mobile analytics

There are DPI service providers such as Phorm (<http://www.phorm.com/>) and Kindsight™ (<http://www.kindsight.net/>), which recently filed a patent on what it calls "character differentiation" technology; there are also DPI hardware firms such as Procera (<http://www.proceranetworks.com/>) that can be enlisted for mobile analytics via DPI analyses. DPI can decompose mobile device origins and behaviors at a very detailed level. DPI provides an awareness of what operating system mobiles are running and where in a network; it can identify what application devices are running, their location, and the type of services being used. This multidimensional intelligence enables mobile marketers to make network decisions to improve the end-user's online experience and offer more relevant ads and content.

Another technology, deep packet capture (DPC), complements DPI. While DPI is designed to enable mobile service providers and marketers to see and act in real-time, DPC solutions are designed to capture all the data passing through a network and store it for more subsequent detailed analysis via machine learning algorithms. DPC is the act of capturing, at full network speed, complete network

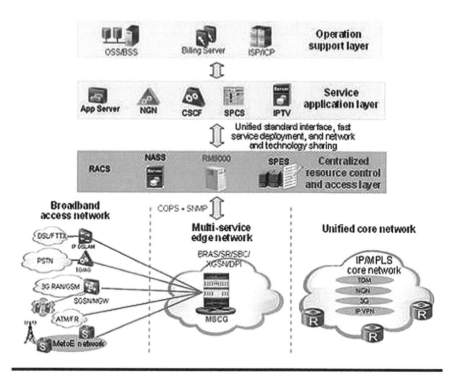

Figure 3.7   DPI layers.

packets, including header and payload data crossing a network with a high traffic rate. Both DPI and DPC offer details about the actual data over a period of time and at multiple points in a network, and for calibrating mobile device behaviors.

3.11 Mobile Marketing Data

Mobiles are not the wave of the future. They are the wave of now. Whether you are defined as a Millenial or someone older but with a substantial income, the mobiles are part of the existence of the majority in major demographic groups. A recent survey from Nielsen (Nielsen.com) did more to state the obvious, especially for brands, companies, and marketers, that the mobiles are being purchased and used in great numbers (Figure 3.8).

While overall mobile penetration stood high overall, those in the 25- to 34-age group showed the greatest proportion of ownership, with 66 percent saying they had a device. In the same age group, eight out of ten purchased a new device every two years. But age is not the only determinant of mobile ownership. Income also plays a significant role. When both age and income are taken into account, older consumers with higher incomes were more likely to have a mobile, a factor brands and marketers must be fully aware of if these are their target markets. For example,

Smartphone Penetration by Age and Income
January 2012

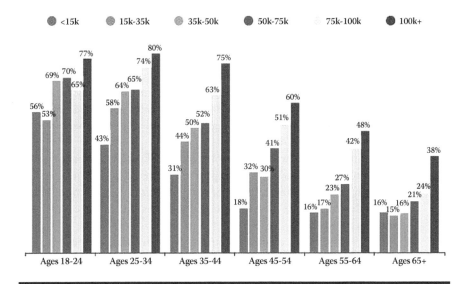

Figure 3.8   Mobiles by age and income.

those 55 to 64 making over $100K a year are almost as likely to have a mobile as those in the 35 to 44 age bracket making $35K to $75K per year.

The demographics should tell the story for any company and brand about these profitable target markets. If these consumers are likely to be carrying a mobile, then brands and their marketing efforts better concentrate on reaching them; it really is that simple. If mobile marketers are hesitating on their mobile strategy, they could be leaving money on the table. If brands feel the urge to do more analysis and dig into these target markets and develop a data mining strategy to reach these profitable consumers.

It is unlikely that brands will uncover a different story, whether they are a business-to-consumer (B2C) or business-to-business (B2B) company. People research everything on-the-go these days. To be more specific, marketers and brands are likely to want to reach these mobile mobs. The simple question for brands, businesses, and marketers: Are they going to meet them there or let their competition do it instead?

3.12 Mobile Data Aggregation Networks

These data aggregation networks—mostly advertising companies and third-party ad exchanges—work like this: the first time a mobile site is visited, they install a cookie or a beacon file, which assigns the visitor's device a unique ID number. Later, when the user visits another site affiliated with the same ad exchange, it can take note of where that mobile was before, where they are now, and their browsing behavior, such as what keywords they used in their mobiles.

This way, over time, these data aggregation and sharing networks can build robust but anonymous, advertising anywhere profiles of mobiles users. The information that these networks gather is anonymous, in the sense that the users are identified by a numbered cookie assigned to their mobile device by these ad exchanges and ad networks; for example, they do not know the names of consumers—only their device behaviors and attributes, as identified by their cookie code number. Following are some of largest mobile data aggregation and ad serving networks:

- TARGUSinfo® (<http://www.targusinfo.com/>): Provides demographics-based data to mobile marketers, enterprises, and websites (Figure 3.9). The network can identify potential prospects at the point of interaction; it can pinpoint a device's precise location and verify names, addresses, emails, and IP addresses in real-time.
- Quantcast (<http://www.quantcast.com/>): A media measurement service for audience statistics for millions of websites and mobiles. They can deliver "look-alike converter" segments of mobiles.
- BrightRoll (<http://www.brightroll.com/>): Provides mobile video advertising services for ad campaigns. They leverage contextual, demographic, geographic, and site-specific targeting methods to help advertisers reach their mobiles' targeting needs.
- RocketFuel (<http://rocketfuel.com/>): Provides mobile media ad campaign management by combining demographic, lifestyle, purchase intent, and social data with their own suite of targeting algorithms, blended analytics, and expert analysis.

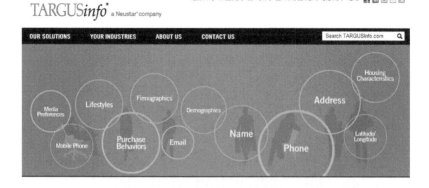

Figure 3.9 TARGUSinfo provides detailed profiles of mobile users.

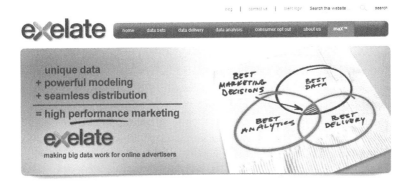

Figure 3.10   Another mobile data aggregator network.

- Turn (<http://www.turn.com/>): Provides end-to-end platform for managing digital advertising; offers a self-service interface, optimization algorithms, real-time analytics, interoperability, and scalable infrastructure.
- DataXu (<http://www.dataxu.com/>): Offers a real-time ad optimization platform for mobile marketers. They start by assimilating client existing data; they then analyze it, looking for patterns of device behavior that help it understand or differentiate levels of intent among target audiences.

Mobile analytics at sites are about using cookies or beacons via ad exchanges and data aggregation networks to customize a visitor's experience and to offer relevant content to them. In addition, data management and data marketplaces such as eXelate (<http://exelate.com/>) allow for mobile analyses to take place in order to get a sense of a consumer's interests (Figure 3.10).

3.13 Mobile Video Data

A recent Allot report (<http://www.allot.com/index.aspx?id=3797&itemID= 83869>) from the bandwidth management firm MobileTrends (<http://www.mobiletrends.ca/>) says video streaming continued to show significant growth with an 88 percent increase each year, and remains the single largest application taking up bandwidth, accounting for 42 percent of mobile bandwidth. Although having limited impact on the total bandwidth, VoIP and instant messaging have gained additional share and continue to be the fastest growing application types with a 114 percent increase. This data seems to be in line with the declining SMS and international voice-call revenue experienced by operators, the report said.

The Allot report also found that Android Market traffic grew by a phenomenal 232 percent, almost four times faster than the Apple App Store^{SM}; YouTube now accounts for 24 percent of total broadband traffic; WhatsApp (<http://www.whatsapp.com/>) now accounts for 18 percent of the IM (instant messaging) total bandwidth, a dramatic increase in popularity from only 3 percent in 2011; and Facebook messenger rose from zero to an astonishing 22 percent of total IM traffic.

The phenomenal growth of OTT VoIP (over-the-top VoIP) and IM represents both a challenge and an opportunity for mobile operators. Intelligent, application-based data pricing is the way forward for operators, allowing them to maximize data revenues based on its true value to millions of mobiles. For the mobile developer and marketer, it also provides a new revenue stream: video mobile data.

3.14 iPhone and Android ID Numbers

Mobile analytics and marketing can take place on the basis of frequency, device type, location, interests, recent activities, preferences, historical behavior, or keywords—all are critical in the triangulated modeling of moving devices. Consumers demonstrate what is useful and valuable to each of them by their actions in search and social sites as well as the interactive mobile world. The role of mobile analytics is to deliver and fulfill—if not *anticipate* users' needs—on any mobile anywhere, at any time. Mobile analytics depends on the triangulation of a mobile, which is transmitting, by its digital unique number, its desires and location.

Universally, all mobiles have unique serial number identifiers, including Apple and Android devices. For the analyst and marketer, these unique numbers of devices are important attributes for behavioral analysis as the numbers are difficult or impossible to delete and can be triangulated via location, content sought, and social links. These unique numbers, which really represent anonymous mobiles, have behaviors that can be modeled in a very precise manner for improved customer service and revenue market growth.

Customized apps can be created by enterprises and mobile marketers with such functions as store locators, proximity coupons, and local real-time ads. The Apple UDID is the most common identifier—it is stamped on all Apple mobiles. It is a combination of forty numbers and letters set to stay with the device forever—stamped because of warranty issues, the UDID has also tremendous mobile marketing applications potentials

The Android ID is the second most common mobile device identifier, and it is embedded by Google on all its mobile operating systems. The ID is set by Google and created when a user first boots up the mobile. Google does allow users to reset their ID numbers, but they also can be used for mobile analytics of these devices via interest- or location-based behaviors.

Originally, these IMEI (international mobile equipment identity) numbers were set up to lock down phones that had been reported stolen or lost; however, their true value lies in mobile analytics and marketing. There is also the IMSI (international mobile subscriber identity) number assigned to all devices; this number is used to route calls and bill users. This footprint will remain the same as long as the user owns that mobile device. These mobile identifiers have important uses for analysts and marketers; in addition, customized apps can be developed to market to unique Apple UDID or Android ID devices—in an anonymous but highly relevant and valuable manner.

A new marketing paradigm is also evolving in which the behaviors of mobile users and their devices are beginning to take place. Targeting on the basis of frequency and recent activities bases on historical behavior is critical in modeling via inductive behavioral analytics for triangulation marketing. Consumers demonstrate what is useful and valuable to each of them by their actions in mobile websites and their apps.

Mobile analytics is done by *multilateration,* also known as hyperbolic positioning. This is the process of locating an object by accurately computing the time difference of arrival (TDOA) of a signal emitted from a mobile device and its triangulation and the synchronization by multiple receivers. A location-based service provider such as Antenna Software (http://www.antennasoftware.com/; Figure 3.11) can be enlisted for customizing mobile apps and sites, and data mining mobiles.

Antenna (<http://www.antennasoftware.com/>) lets mobiles take their shopping cart with them wherever they go. Antenna Software provides an intelligently designed, user-friendly mobile app that consumers can access at their convenience to securely browse and buy products right from their digital devices. With personalized alerts, coupons, discounts, deals, and other services, their "Mobile Shopping"

Figure 3.11   Location-based software.

component allows for mobiles to keep those consumers actively engaged who are more likely to make future purchases.

With mobility, the data miner and marketer can take that relationship to the ultimate level by providing an interactive, personalized shopping experience to mobiles—wherever and whenever they are in the mood to shop. For example, the Mobile Shopping app by Antenna Software (<http://www.antennasoftware.com/>) allows mobiles to receive mobile coupons, point-of-sale promotions, and specials—all customized by the retailer. The app also supports the location of retail items while shopping in the physical store of the merchant. The Antenna app also supports the use of the mobile's camera—to capture product barcode for pricing, availability, and more information.

The Antenna app can also be used to search for items, view products and images, or add to a shipping cart, add to a mobile's wish list, and finally check out. The Antenna app can also be used to find nearby stores and get directions from current location using the mobile's GPS. Many of today's mobiles typically determine location by analyzing a mix of wireless signals from GPS, cell towers, and Wi-Fi hot spots. Wi-Fi signals are considered the most accurate, as they work indoors and in urban areas; they also can be retrieved quickly and are accurate to within twenty to thirty meters.

That level of precision is important to mobile marketers and app developers, who want to be able to send time-sensitive targeted ads, offers, and coupons to consumers when they are in their mall or close to their stores. Many different standards have been established for how mobile phone networks are supposed to communicate and coexist with each other. Currently, the most popular mobile device communication standards are the Global System for Mobile Communications (GSM) and the Code Division Multiple Access (CDMA).

However, a new high-speed standard called Long Term Evolution (LTE) is being introduced in populous areas. To counteract unauthorized use of their networks, mobile phone providers have introduced security measures that do not allow a mobile to receive and transmit calls without certain valid data. This data is usually stored on small microchips called Subscriber Identity Modules (SIM) cards, and the data they carry is known as the service-subscriber key.

The technical name for this data is the International Mobile Subscriber Identity (IMSI). This is used to gather information about the mobile country code, mobile network code, and mobile station identification number of the device. It also carries an authentication key. SIM cards are most commonly used in mobiles that are intended for GSM networks. Many devices built for CDMA networks do not use removable cards, and instead embed the security information permanently in memory. However, trans-international devices that can handle both communication protocols include SIM cards; by sending and receiving radio signals through a vast interconnected network, mobiles can make calls almost anywhere in developed countries, over distances of thousands of miles.

3.15 Blocking Browsers

Amid swirling discussion on how tech companies handle privacy issues, Google became the latest Internet giant to support adding a do-not-track button to its Web browser. No time frame was set for changing the Chrome and its Android browser to include a do-not-track feature, which would prevent companies from using information gleaned from a user's Web history to deliver tailored advertising. Companies that have adopted the standard also have agreed not to collect data for use in credit, employment, health care, or insurance decisions.

Google has implemented a "Do Not Track" header in a consistent and meaningful way that offers users a choice and clearly explains the browser controls. Google joins Mozilla, which added a do-not-track button to its Firefox browser last year, and Microsoft, which brought it to Internet Explorer a few months later. The next version of Apple's operating system, called Mountain Lion, includes a do-not-track feature in the Safari browser released to developers in October 2012. This blocking feature of browsers will limit the collection of mobile data.

Google's move came in response to the Obama administration's call on lawmakers to create a "privacy bill of rights," designed to give consumers more control over how their data is collected, stored, and shared. The privacy framework, which was developed over two years by the Federal Trade Commission (FTC.gov) and resulted in a sixty-two-page document, sets forth privacy principles that regulators say companies should uphold. The goal is a policy that would ensure that consumers can control what data companies collect about them, understand how it can be used, and prevent companies from using it in ways they did not explicitly agree to.

> "Even though we live in a world in which we share personal information more freely than in the past, we must reject the conclusion that privacy is an outmoded value," President Obama wrote in a letter accompanying the report. "It has been at the heart of our democracy from its inception, and we need it now more than ever."

Privacy advocates say that a do-not-track button, while a positive step, will not ensure Internet users' privacy by itself. A separate effort by the advertising industry will allow people to opt out of being tracked by clicking a virtual off-switch within advertisements, but that project is not linked to the browser buttons. For that reason, a person who opts out of sharing personal data using one method may still be tracked by other means—in a way that could be confusing. "I would like to see us move to a place where it isn't so burdensome for consumers, " said the FTC (FTC. gov) commissioner.

The battle now moves to Congress, where lawmakers are expected to hold discussions about privacy legislation. Privacy advocates say they are concerned that their voices will be drowned out by the large companies that depend on targeted advertising. The real question is how much influence companies such as Google, Microsoft, Yahoo!, and Facebook will have in their inevitable attempts to "water down" the rules that are implemented, and render them essentially meaningless.

3.16 Mobile Voice Recognition Data

Increasingly, users will be talking to their mobiles, asking them for directions to places, stores, or other friends. The most prominent device that supports this is, of course, the iPhone® 4S and its AI app Siri® that lets users send messages, schedule meetings, place phone calls, and more. Users can ask Siri to do things, and the mobile understands what they say, knows what it means, and even talks back. Mobile analysts and marketers need to leverage these voice and facial recognition technologies to facilitate interactions and increase sales and consumer loyalty. In addition, voice recognition technologies from such firms as Vlingo (<http://www.vlingo.com/>) and PhoneTag® (<http://www.phonetag.com/>) can enable consumers to converse with their devices, instructing them to perform searches for them that can be delivered with targeted content and advertisements.

Vlingo, for example, combines voice recognition technology to quickly connect with people, businesses, and activities; it can read incoming text and email messages aloud (Figure 3.12). The Vlingo system allows users to say anything to their mobile and still be recognized properly. Accuracy is greatly improved for tasking devices—and is better than humans on many tasks; any application can be speech enabled. This is accomplished through advanced adaptation techniques that the software learns over time and what users are likely to say to different applications.

Central to these techniques is a technology called Hierarchical Language Models (HLMs) that allows Vlingo to scale its learning algorithms to millions of

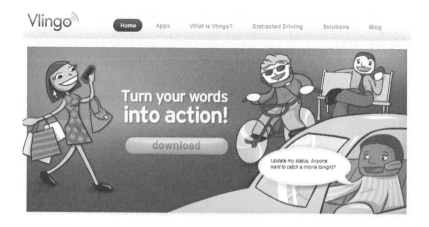

Figure 3.12   Vlingo can be downloaded to mobiles.

vocabulary words and millions of users. To achieve unprecedented accuracy at massive scales, Vlingo technology uses HLM-based speech recognition that replaces grammars and statistical language models with a very large vocabulary using millions of words; these HLMs are based on well-defined statistical models to predict what words users are likely to say and how words are grouped together. For example, "Let's meet at ___" is likely to be followed by something like "1 pm" or the name of a place.

While there are no hard constraints, the Vlingo models are able to take into account what this and other users have spoken in their software's textbox in a particular application, and therefore improve with usage. Unlike previous generations of statistical language models, the new HLM technology being developed by Vlingo scales to tasks requiring the modeling of massive amounts of possible words, such as open Web search, directory assistance, navigation, or other tasks where users are likely to use any of a very large number of words.

PhoneTag can also convert voicemail to text and deliver it via email and/or as a text message (Figure 3.13). PhoneTag works with all major U.S. carriers and networks, including: AT&T, Alltel wireless (<http://www.alltelwireless.com/>), Cincinnati Bell[SM] (http://www.cincinnatibell.com/), Sprint, Skype, T-Mobile, Verizon, Virgin mobile® (<http://www.virginmobileusa.com/.), and more. PhoneTag can be used to unify all devices,, whether mobile or desktop into one unlimited voicemail box. With PhoneTag, users can read every voicemail that comes into any of their mobiles' phone numbers from anywhere.

Figure 3.13   With three easy steps, PhoneTag® converts voice to text.

Yet another voice recognition firm is Microsoft Tellme® (<http://www.micro-soft.com/en-us/tellme/>), which provides instruction technology so that people can use their mobiles to get the information they need. It combines Internet data and a voice interface, and allows users to simply say what they want and get it. Instead of fiddling with buttons and memorizing keystrokes, users can *tell* their mobiles what they want. Some of the services running on the Tellme platform include business search on 411 and information search on 1-800-555-TELL.

Clearly, the best voice recognition technology is Siri, which understands nat-ural language coupled with it natural user interface. There are Siri-like apps for Android, such as Speaktoit® Assistant (<http://www.microsoft.com/en-us/tellme/>; Figure 3.14) and software developers are waiting for Apple to open Siri's application program interface to the outside world so they can begin creating complementary apps. Siri and other new applications understand natural speech.

Siri and Speaktoit are amalgams of speech-recognition algorithms with a hint of artificial intelligence (AI) and natural language processing blended in; that keeps them getting smarter over time as these assistants get to know the user better. Voice interface simplifies the way people communicate with services and their mobiles; these personal assistants with their voice interfaces will become a new important method by which marketers and brands communicate with consumers. These voice recognition apps are making computers behave more like humans instead of the other way around, and they represent a new way to aggregate mobile data.

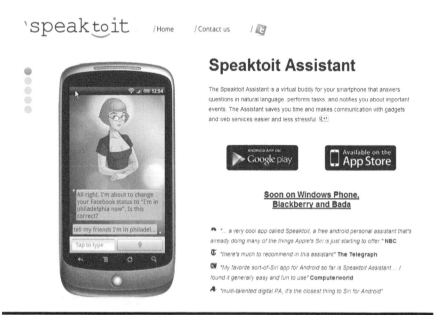

Figure 3.14   The speaking assistant for Android.

NuanceCommunications (<http://www.nuance.com/>) recently introduced long-awaited mobile versions of its popular Dragon Medical Mobile Recorder and Dragon Medical Search speech recognition software, in the form of an Apple app. This is important because speech recognition can speed up data capture on many platforms, perhaps more so with mobile. This is important because speech is three times faster than typing on a standard keyboard, but it may be five to six times faster than typing on a touch-screen or tiny buttons of a mobile!

Nuance is looking to shift much of the processing of data from mobiles to the cloud in order to offer greater computing power and to safeguard sensitive healthcare information in case a device gets lost or stolen. The other benefit of cloud processing is that users can keep the same profile no matter how or where they access the mobile data and application. Interestingly, privacy laws in many European countries generally do not allow Nuance to host healthcare information in the cloud, so institutions often make use of private, in-house clouds that Nuance is able to update to.

Some Canadian institutions also go this route because they are afraid of being subject to U.S. Patriot Act reporting requirements if they send data across the border. Some larger healthcare providers in the United States also opt for private clouds and even host clouds for other organizations in their service areas. All Nuance internally developed mobile apps are free to download, but some require the purchase of host software on the server end.

Clinical language understanding (CLU) is in some existing Nuance products but has not been released for third-party developers yet. CLU essentially is a means of helping computers make sense of all the unstructured data coming into electronic health records and other healthcare software. CLU is basic fact extraction: clinical data tagging, procedures, problem lists, medications, and allergies. So whether it is speech-recognized text that is coming in and then being parsed out and structured for the user or even actual text documents to begin with, there are a variety of different ways to get that data into a structured format. Demand for text-to-speech services is growing as well; Nuance reports seeing the acceptance or the expectation for that to occur in a consumer environment. Right now, Nuance mobile apps are only designed for Apple devices.

In the future, machine interaction will become more natural and human centered, and mobiles will be able to understand body language, expressions, voice tones, and complex emotional content. Siri is a good step in this direction in that it carves off a piece of "fuzzy" interaction that humans do; it builds the AI around a very functional role—that of an assistant with expected behaviors, helping make appointments, reminders, turning things on or off, providing directions, and taking dictation. Voice is only the first step; gesture and facial recognition will follow, as will the capability of mobiles to be sensual—thus enabling consumers to feel textures or edges of products.

3.17 Mobile Data Assurance Service

Polystar (<http://www.polystar.com>), a supplier of service assurance, network monitoring, and test solutions, has unveiled its Data Service Assurance Solution. The new solution delivers visibility into subscribers' behavior, device usage, service quality, and performance of all mobile data networks, including LTE. The company has dedicated major R&D resources to make sure they have a state-of-the-art platform for mobile data monitoring. The company said that it has rebuilt its Mobile Data Monitoring solution to address the challenges of an ever-changing mobile data environment.

The complex ecosystem, characterized by a wide range of mobile interactive services, forces network operators and service providers to proactively manage the outrageous traffic volumes and adjust their daily operations correspondingly. Polystar added that the new scalable platform will handle today's bandwidth demands, as well as support future data explosion and service migration from the circuit to the packet world. With the introduction of new releases, the enhanced system architecture, specifically developed to handle mobile data, becomes generally available.

The company provides network and service assurance solutions allowing for full visibility of networks and service performance. This visibility combines with an easy, actionable path down to the root cause of a problem. A unique feature of this solution is the ability to analyze the actual payload, thus offering a totally new level of insight into service quality. It enables end-to-end monitoring across all types of network technologies, as well as exceptional insight into the individual services carried over these networks.

Network and service performance is measured by various performance indicators and can be applied to any specific measurement or network event. Real-time alarm generation on service degradation facilitates proactive and rapid problem resolution. In addition, network and service indicators are displayed on a fully customizable dashboard with drill-down capabilities.

Polystar also recently launched the test tool Solver, which includes VoLTE (Voice over long term evolution), which simulates EPC (Evolved Packet Core) traffic over the SGi (Small Group instructions) interface to the IMS network (P-CSCF (Proxy-call session control function)). Via efficient VoLTE traffic simulation, the functionality, quality, reliability, and ability of the IMS (IP multimedia sub-system) network can be tested and verified. Polystar designs, develops, delivers, and supports systems that increase the quality, revenue, and customer satisfaction of mobile data services.

3.18 Mobile Facial Recognition

A new biometrics feature has been added to mobiles, allowing an enterprise to positively identify employees reporting to work. FaceFront Biometrics, from

ExakTime (<http://www.exaktime.com/>), allows their company's PocketClock/GPSmobiletimeclock, on any mobile, to turn into a photo-verification device. Users simply use the camera lens of the device to snap a picture of their face and FaceFront will allow bookkeepers to match an employee's field photo to a file photo, thereby verifying the employee. The FaceFront Biometrics app was originally developed with contractors in mind; however, this same face recognition technology can be used to recognize consumers via mobile data.

There are some features in the Android operating system that reach into the futuristic-sounding realm in order to catch people's attention. Android Beam™ and FaceUnlock can recognize the user and use face recognition technology to unlock mobiles. Once a user loads a photo of their face onto the Android mobile, the Face Unlock feature can use that photo to compare whoever is trying to unlock the mobile. If it is a match, then the device is enabled; it can tell if someone is holding up a picture and trying fool it.

Then there is the Android Beam, a feature that makes use of the near-field communication (NFC) chip that will be on more and more mobiles in the future. The Android Beam works when two Google mobiles having newer versions of Android OS with the NFC technology are tapped together, allowing for the instant transfer of apps, contacts, sites, and videos between the devices.

SensibleVision (<http://www.sensiblevision.com/>; Figure 3.15) has a patent-pending mobile version of its face recognition technology that allows near-perfect

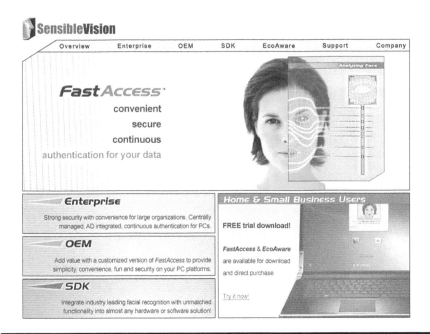

Figure 3.15   SensibleVision for recognizing consumers.

photo and video rejection. The company has released a new product that offers convenient and secure face recognition for Android and Apple mobiles. According to the company, their Faced.me app uses proprietary technology to recognize a person's face in less than one second by identifying different facial points. It then matches those faces to the social networking accounts for that user, allowing users to friend them on Facebook, follow them on Twitter, or connect on LinkedIn. This type of technology enhances the value and functionality of mobile data.

3.19 Mobile Wallet Data

Mobile payments will become ubiquitous; unlike previous failed efforts to allow consumers to pay by tapping their mobile phones at points of sale instead of swiping credit cards, the new push to expand mobile payments is focused on transforming shopping. Previous efforts to take mobile payments to the mainstream failed because shoppers saw little benefit in trading a payment card for a mobile; and because merchants have been loathe to invest in new devices to accept payments.

By contrast, a mobile model is evolving by companies such as Isis (<http://www.paywithisis.com/>) that would allow consumers to consolidate credit cards and customer loyalty cards in a wallet app that they can use to unlock deals when they enter a store (Figure 3.16).

The credit card form is limited; however, there is no such limit with a mobile wallet and its software mechanism. With three of four U.S. consumers already using mobiles to help guide in-store purchasing decisions, price comparisons, and product searching, Isis emphasizes its mobile payment system's effect on the overall shopping experience. Isis is a joint venture of AT&T Mobility, T-Mobile USA, and

Figure 3.16 The Isis mobile wallet.

Verizon Wireless with its payment system. The company is essentially not shifting the way consumers pay; rather, they are bringing to consumers an opportunity and ease of use to transform the way they shop.

Isis is addressing the problem of merchant adoption via partnerships with VeriFone® (<http://www.verifone.com/>), Ingenico® (<http://www.ingenico.com/en/>), ViVOtech® (<http://www.vivotech.com/>), and Equinox Payments system (<http://www.equinoxpayments.com/>), which will integrate and support the Isis Mobile Commerce Application in its current and future product lines. The agreements with leading payment systems providers will help accelerate the wide-scale introduction and adoption of mobile commerce. These leading suppliers of payment terminals have agreed to begin enabling the NFC technology to power mobile payments in all the machines they provide. Merchants who have not moved to the new terminals by 2015 will assume increased liability for fraudulent credit card purchases at their stores.

Major manufacturers of Android mobiles, including Nokia, Sony, and LG, have already incorporated NFC in their devices. Already over 100 million NFC-enabled mobiles exist, and mobile payments using the Isis model are actually safer than credit card transactions. This is because the data on a credit card's magnetic strip can be "skimmed," thus allowing criminals to create a pirated card.

However, the security code attached to mobile payments using the Isis system will change because the data is not static, so stealing it does not do the criminal any good. Also, a lost or stolen wallet app, unlike a lost or stolen "leather wallet," can be deactivated with one call to the mobile provider. In addition, if a mobile's battery dies, the user would not be left without access to mobile payments because the NFC reader can access the app with its own juice; it also can be blocked in order to ensure that the wallet app is always protected via a PIN number.

This type of mobile wallet payment data can provide a treasure of information and consumer intelligence to merchants and marketers; it can provide vital patterns about what mobiles are buying—when, where, and why. This wallet data can help discover important merchant intelligence about the price points and marketing efforts that are creating the most sales about their products and services. This data provides vital metrics and insights in order to build new product features, inform the product roadmap, and inform merchants and brands about product and sales performances.

3.20 Mobile Digital Fingerprinting Data

Today, mobiles are the targets of analysts and marketers using website cookies, apps, and now a new technology known as *digital fingerprinting* that allows for the modeling of machines, not user behaviors. Digital fingerprinting tracks the behaviors of mobiles, not humans; patented technology makes it possible for marketing to any and all mobiles based on such attributes as the device's operating system, fonts, color settings, and hundreds of other machine-specific settings. It takes mobile data to a whole new level of efficiency.

The technology relies on "device identification." New firms such as BlueCava (<http://www.bluecava.com/privacy-policy/>) strip out all personally identifiable information (PII), so that the tracking and targeting are based on the features of mobiles, not the people. Digital fingerprinting technology uses a new way to aggregate consumer intelligence around a universal identifier; BlueCava uses their proprietary token, which is persistent; however, because the technology is profiling mobiles, there are no privacy concerns for enterprises, brands, marketers, or consumers.

Digital fingerprinting creates a single "reputation profile" based on the mobile behaviors, shopping habits, and other machine-specific attributes, features, and settings. BlueCava is racing to collect digital fingerprints from every type of digital device—including all the mobiles in the world—to perform device identification for the modeling of their behaviors and targeted marketing. The company can uniquely identify any mobile with more than 99 percent accuracy around the globe!

Thus far, BlueCava has identified 200 million digital devices with the hope that by the end of next year, they will have cataloged one billion of the world's estimated 10 billion devices. Digital fingerprinting is anonymous, very robust, and more accurate than traditional Internet mechanisms such as cookies, beacons, and apps. Fingerprinting involves capturing and modeling devices based on several parameters, including but not limited to hundreds of settings and attributes such as the following:

- *Fonts*: It looks to see what fonts were installed and used on a mobile.
- *Screen size*: It looks to see what screen size and the color setting of the mobile.
- *Browser plug-ins*: It looks and maps what optional software a mobile has, such as apps.
- *Time stamp*: It compares the time a device logs on to a server, down to the millisecond.
- *User agent*: Similar to cookies, it identifies what type of operating system the mobile is using.
- *User ID*: It assigns a unique token that can be used to track all activities for that mobile.

BlueCava device identification technology creates an electronic fingerprint based on the unique characteristics of any mobile. This means that no two digital devices are seen as identical; as is the case with human fingerprints, digital fingerprints are unique to each machine. Being able to identify and differentiate devices represents a powerful tool that can be used for authentication, auditing, access control, licensing, and fraud detection; but most importantly, the technology can be used to achieve a new level of precision and targeting for the data mining of mobile devices. The BlueCava device identification platform provides a unique combination of three characteristics essential to mobile tracking:

1. *Uniqueness:* Their device fingerprinting is based on dozens of component types and attributes plus values, which guarantees the uniqueness of a mobile.
2. *Tolerance:* Their device identification algorithms are resilient to changes in physical mobiles and their configuration.
3. *Integrity:* Their obfuscation, hashing, encryptions, and randomization work in unison to provide integrity to the secure device fingerprint.

There are two kinds of fingerprint clients: physical device fingerprint clients and Web fingerprint clients. Physical device fingerprint clients are installed and run on the actual mobile device. The physical device client can be packaged within apps or can be downloaded and installed as a stand-alone. Physical device fingerprint clients are available on a variety of platforms, including Apple, Android, BlackBerry, and other mobiles. The Web fingerprint client, on the other hand, is run from within a browser and can be invoked automatically from a webpage, with no additional download required. Both fingerprints are passed to the mobile lookup service for resolution into a persistent device token. Aside from BlueCava, there are several other firms offering fingerprinting technology, including iovation® (<http://www.iovation.com/>), 41ˢᵗ Parameter® (<http://www.the41.com/>), and Imperium® (<http://www.imperium.com/>). All of them, just like BlueCava, are principally using digital fingerprinting technology for security and fraud detection, but are also amenable to mobile analytics and marketing.

iovation offers what it calls "Real IP" for identifying a user's IP address and geo-location. Location details help users of iovation's service and software pinpoint mobiles anywhere on the planet. Their Real IP makes it easy to triangulate and target the actual location of any mobile device. The iovation Real IP tells the mobile analyst where visitors to sites are coming from. Geo-location data includes country, stated and real IP addressed, latitude, and longitude. Real IP verification works in real-time with IP verification in milliseconds. Mobile analytics can set the usage parameters and extract business rules, which can be developed via decision tree analyses. Their "ReputationManager360" exposes the behaviors and reputations of every mobile that interacts with a site.

41ˢᵗ Parameter offers what it calls "DeviceInsight," that enables websites to "converse" with all transacting stationary or mobiles. Through this nonobtrusive, automatically conducted conversation, a digital fingerprint for the mobile is created, which can then be used to match devices to log-ins or transactions. DeviceInsight requires no user involvement, hardware deployment, or disruption to the user experience.

41ˢᵗ Parameter offers an open API of DeviceInsight that allows it to be easily integrated into existing applications, operating with any device connecting to a website, including mobiles via any browser connection. Web and mobile cookies are capable of profiling, on average, about 70 percent of all devices; DeviceInsight can generate device-based proxies for virtually 100 percent of all machines without infringing on consumer privacy.

Based on patented technology, "DeviceInsight for Marketing" generates a digital fingerprint so granular that it can serve as a functionally superior replacement for cookies or Flash Local Shared Objects (LSOs). Requiring no user involvement, hardware deployment, or change to the user experience, it does not write anything to the consumer digital devices. The technology does not leave any residue in the form of a cookie, Flash LSO, or any other object, thereby making it an effective and transparent solution for mobile analytics—even in countries with the strictest personal privacy regulations.

To further differentiate between devices, especially mobile ones that have fewer attributes available to generate a digital fingerprint, 41st Parameter has the patented Time Differential Linking (TDL) software that increases the uniqueness of device fingerprints by up to 43.7 percent compared to the accuracy of fingerprinting alone. This level of differentiation allows mobile analytics to target more accurately, with up to 100 percent longer duration than with traditional JavaScript cookies. Their technology is packaged as both a Web/HTML API and a mobile app SDK for Apple iOS.

Imperium, another digital fingerprint firm, offers three products: RelevantID for mobile analytics and data certification, Verity for identity validation, and RelevantView for market research. As with the other fingerprinting companies, Imperium is also involved in the dual areas of fraud detection and real-time analytics. The RelevantID digital fingerprinting technology goes beyond traditional cookie methods to create a new approach to ensuring and certifying that the behavioral data collected is reliable and predictive of mobile behaviors.

RelevantID works via a combination of watermarking and digital fingerprinting. Digital fingerprinting is the process of collecting more than sixty data points about a user's mobile. These data points are then processed by their proprietary algorithm to produce a unique digital fingerprint. RelevantID is Imperium proprietary technology and is consistent with U.S. privacy and European data protection laws, in support of marketing machine-to-machine, and is highly precise for enterprises and relevant to consumers.

When a user accesses a service, RelevantID can identify the mobile and collect a large number of data points about its attributes and settings. The information gathered is put through deterministic algorithms to create a unique digital fingerprint of each mobile device. The Imperium process is invisible to the user and does not interfere with the user experience. Once fingerprinted, a device is open to mobile analytics. For example, the fingerprinted devices can be clustered into different categories using self-organizing map (SOM) neural network software, or they can be segmented into distinct groups based on machine learning decision-tree business rules. These modeling technologies and software are discussed in Chapter 5.

Digital fingerprinting is the future of mobile analytics, largely because it is invisible, tough to fend off, nonintrusive, more accurate than cookies, semi-permanent, and it works with *all* types of digital devices. A typical mobile device broadcasts hundreds of details about itself when it connects to a site or server in an industry estimated to exceed $23 billion in the United States alone. Each digital device has a different clock setting, fonts, software, operating system, apps, and many other characteristics that make it unique.

These digital fingerprinting companies do not collect individuals' names; they only collect the digital details of devices. BlueCava can embed its fingerprinting technologies in mobile sites, apps, and mobiles. This can also include the matching and merging of mobile apps and the attributes and mobile device behaviors captured from websites. For example, Mobext (Mobext.com), a mobile advertising unit of Havas SA, is already testing BlueCava fingerprinting technology on mobiles.

The importance of digital fingerprinting will increase in the future as more and more mobiles are being manufactured with embedded processors from Intel and other microprocessor manufacturers. Intel® (<http://www.intel.com/>) sees digital advertising signage as a new growth market for their Atom™ chips. LG Electronics (<http://www.lg.com/us>), for example, is using Intel Atom chips in signs that will recognize the age, gender, and other characteristics of passersby and change the advertising pitch accordingly. ARM® Holdings (<http://www.arm.com/>) is also marketing smart chips to Texas Instruments™ (<http://www.ti.com/>), Qualcomm® (<http://www.qualcomm.com/>), and Marvell® (<http://www.marvell.com/>), who are embedding them in a diverse group of mobiles.

3.21 Capturing Mobile Data

To meet the growing global market demand for its mobile data capture solution naturalFORMS®, ExpeData® (http://www.expedata.net/; Figure 3.17) is providing

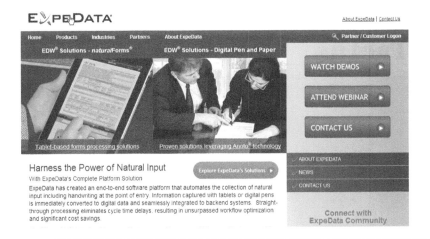

Figure 3.17 **Capturing mobile data.**

its software in nine languages, including Dutch, English, French, German, Italian, Portuguese, Russian, Spanish, and Swedish.

With their new localization enhancements, users can access their menus and keyboards and view handwriting interpretation in their native language, further enhancing the overall flexibility and usability of the application. naturalFORMS allows users to easily and efficiently complete business forms on Apple and Android tablets using natural input such as handwriting.

Input captured on the mobile is immediately interpreted and converted to digital data with instant verification and feedback if form data is incomplete or does not adhere to enterprise rules. Straight-through processing allows high-quality data to flow seamlessly into back-end systems, providing unsurpassed workflow optimization. naturalFORMS provides the best handwriting interpretation accuracy on the market by leveraging MyScript technology from VisionObjects (<http://www.visionobjects.com/>), a provider of handwriting recognition technology that is recognized as one of the most advanced and accurate handwriting recognition technologies available, with more than 20 million users worldwide.

3.22 Mobile Haptics Data

A new technology known as *haptics* will add tactile dimensions to mobiles so that consumers will be able to feel the textures of products they are shopping for before they purchase them. Haptics software and digital screen start-ups such as Pacinian purchased by Synaptics.com, Senseg, and Tactus Technology (<http://www.tactustechnology.com/>; Figure 3.18) are racing to add tactile sensations to mobiles—and add a new level of consumer interaction and shopper engagement—and the promise to fill the basic human desire to *feel* things. Mobile analytics will complete the shopping experience—not only will apps locate that pair of jeans or purse the

Figure 3.18   Tactus Technology brings new levels of interaction to mobile data.

consumer was looking for, but they will also be able to do price comparisons, perform different configurations of colors and styles, and they will also be able to touch and feel them prior to purchasing them—all from their mobile.

Millions of Nokia, Samsung, and LG devices have already been preloaded with haptics software made by Immersion for gaming and marketing apps. Touchscreens and -surfaces are quickly emerging as the preferred interface for new mobiles. Pacinian has developed an interface for creating the sensation of pushing a button with no mechanical parts. The Senseng coated screens can produce feelings such as vibrations, clicks, and textures. TactusTechnology has patents on interactive pop-up screens, allowing consumers to feel the edges of letters, numbers, symbols, knobs, instrument strings—and most importantly, the textures of products that can pop up on their mobiles—upon the request of their human owners.

Immersion (<http://www.immersion.com/>) conducted a study employing quantitative and qualitative techniques to evaluate consumer attitudes about haptics in mobiles. The controls included devices that used industry-standard haptic techniques and mobiles that did not have any haptic features. The key findings include the following:

- The vast majority (90 percent) of respondents said they preferred mobiles that offer haptics, compared to standard or no-haptics alternatives. Very few respondents said they did not care one way or the other about the feature. Among those who preferred the alternatives, a greater number preferred standard haptics rather than non-haptics handsets.
- The study found that consumers have clear and measurable preferences for haptic-enabled apps. The study gave participants the opportunity to measure their preferences for haptic implementations in two representative games: Frozen Bubbles and Pinball. Overall, they preferred the haptic versions of each game compared to the controls. Further, the sophistication of the haptic implementation had a significant impact on their responses; their preference for haptic options increased substantially, and across all performance parameters evaluated, when the implementation was customized to deliver specific rather than generalized tactile effects.
- Haptics improve the user's application experience, and haptic effects produce an even better experience. The participants were given a chance to evaluate a haptic-enabled typing application that produces the sensation that the buttons move when touched. The participants said they preferred the option offering haptic effects.
- Next-generation apps are perceived by consumers to have high value. The study participants were introduced to a new concept for haptics, called "expressive alerts," which gives users the capability to use haptics to personalize the various types of messaging alerts and other notifications they receive on their mobiles. The suggested features had extremely high appeal and

almost all participants said they were likely to use the feature if it becomes available on their mobile.

■ The study found that the ability to customize and personalize a phone with haptics was very important to consumers and therefore critical to the success of the technology in mobiles. After testing the typing application, almost all participants expressed a desire to be able to increase or decrease the intensity of the touch sensation. The ability to customize alerts with different haptic sensations—especially to differentiate between personal and business messages—was important to the vast majority of participants.

3.23 Mobile Mad Men

The characters on the hit television show *Mad Men* and the real-life mobile operators of today are pretty much the same. Both could enjoy cigars and martinis while business runs as usual, living as kings of their respective business worlds. Both Madison Avenue and the mobile carrier world are getting turned on their heads by the likes of Google and Facebook. Both industries are reinventing themselves, and the good news is that they actually need each other to help them survive and thrive. *Mad Men's* main character, Don Draper, famously said, "Advertising is about one thing: happiness." Witness the purchase of California-based Amobee by the mobile carrier SingTel (<http://info.singtel.com/>) for $321 million in an all-cash deal. Now that is happiness for Amobee!

This transaction is one of the largest in the mobile ad space behind the AdMob sale to Google and the Quattro acquisition by Apple. It also signals a major chapter in the carrier world—they are fighting back against the over-the-top plays by Google and Facebook. The carriers want their fair share of the media pie, and realize that now is the time to act. The big question is: Can these telcos add the advertising expertise to create a great value proposition for both brands and consumers that makes them both happy? It is a $22 billion question, or roughly the estimated market size of combining deals, geo-fence mobile marketing, and location-based services.

The world's largest carriers—AT&T, Telefonica (<http://www.telefonica.com/>), SingTel, and others—have all created new business units focused on the delivery of digital advertising and mobile payments. Why? They have all come to recognize their uniquely valuable assets of mobile data for advertisers: they can locate a user anytime and anywhere, they have a trusted relationship with their subscribers, and they can enable a closed-loop, real-world transaction. Most urgently, they are realizing that media and payments represent the best opportunity for growth in their own business in the face of rising infrastructure costs, churn, and declining margins.

For consumers, the proposition could be compelling. From enabling hyper-local offers and the ability to pay with their mobiles, consumers could finally begin to receive media that is truly a valuable service and not intrusive based on the data mining of mobile devices. Consider an opt-in world where consumers pre-select the categories and types of offers they want to receive, such as getting an alert from their favorite store about a sale when they are nearby. With the ability to link their credit card and pay with their mobile, these and many other services are headed that way—and if done right, could transform their mobiles into highly personal and invaluable instruments of commerce.

For advertisers, the proposition is even better. While mobile advertising is taking off with performance-based models, brands are still struggling with mobile—look no further than the decline of the Apple brand-centric iAd that had overly complicated pricing. Direct advertising deals are still nascent as proprietary apps often do not achieve scale and only work when the app is on. By focusing on mobile advertising, carriers can bring brands real relevancy and the reach that has been missing from other mobile initiatives.

Tailored offers triggered when a consumer is near a store without an app is a hugely powerful offering from the carriers. Combine it with consumer preferences and a host of anonymous mobile data about subscribers, and brands can finally get a tool that they can use at every point of contact. This "Mobile *Mad Men*" scenario focuses on the digital purchase funnel that works for driving real-world commerce. Mobile ad men already using this approach are seeing results incomparable to any other medium—as high as 65 percent purchase rates for proximity-based mobile marketing.

And there is real scale: The carriers cited above represent over a billion consumers in thirty-plus countries. In markets such as the United Kingdom, all the major operators are trying to collaborate to make it easy for brands to reach the entire audience of a country with a single ad buy. The potential is substantial: according to analysts; the daily deals market will be $4.2 billion over the next four years (BIA/Kelsey;   <http://www.biakelsey.com/company/press-releases/110913-biake-lsey-revises-deals-forecast-upward-slightly.asp>), while mobile proximity marketing could generate another $6 billion (Borrell Associates; <http://www.borrellassoci-ates.com/component/virtuemart/?page=shop.product_details&flypage=garden_flypage.tpl&product_id=789>) and location-based services $12 billion (Juniper; <http://juniperresearch.com/viewpressrelease.php?pr=180>).

It is the SoLoMo concoction, which stands for social, local, and mobile. While it is a ridiculous acronym, the SoLoMo cocktail, when mixed together using the right recipe, is giving operators visions of their own large, tasty slice of pie via a combination of different operators to unlock this revenue that is to add digital advertising expertise to their "DNA," thus bringing breakthrough opportunities to advertisers—offerings that they cannot get anywhere else. For example, here is something that operators are in a unique position to offer: the ability to deliver a targeted impression on the Web, then possibly on Wi-Fi or in email, followed by a tailored alert when the consumer is near the store without an app.

In addition to transactions such as Amobee, these companies are recruiting heavily from the digital technology companies and ad agencies. They are also opening offices in Silicon Valley and working closely with companies such as Bubble

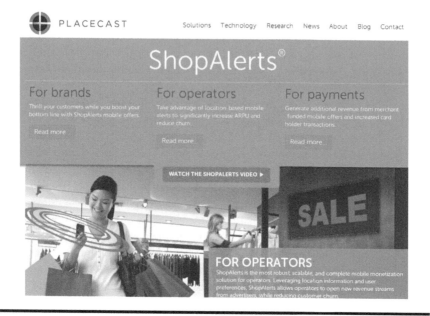

Figure 3.19 Location-based alerts via Placecast.

Motion (<http://www.bubblemotion.com/>) and Placecast (<http://placecast.net/>; Figure 3.19) that enable operators to get a slice of that $22-billion pie, all through a differentiated media offering. The carriers can really succeed here because there is so much more that marketers want from mobile ads that Google, Facebook, and Apple will not be able to offer.

However, carriers must create a differentiated offering based on their unique attributes of user data, real-time location, direct consumer relationships, and the ability to close the loop with transactions, and not just try to replicate the current ecosystem. From APIs for developers to providing rich location context and user data for targeting on both the Web and on mobile, carriers could deliver a consumer all the way down the purchase funnel—from awareness to a real-world transaction. It has become clear that brands and agencies want more than what is there today, and carriers are uniquely capable of unlocking mobile marketing at effective scale. Whether this succeeds or not will simply come down to how fast operators embrace their future. If the industry moves toward SingTel and Telefonica, then the former kings of mobile data can continue to earn a really big piece of the pie.

3.24 Mobile Data Privacy Policies

A mobile analytics privacy notification policy discloses some or all of the ways that enterprises, developers, brands, and marketers gather, use, disclose, and manage a customer's mobile data. The exact contents of a privacy policy will depend on the applicable law and may need to address the requirements of multiple countries or jurisdictions. While there is no universal guidance for the content of specific privacy policies, it is critical that mobile analysts and marketers ensure that such privacy policies and notifications are developed and prominently displayed to consumer mobiles. Failure to do so can doom a mobile analytics project and destroy a company—witness what happened to Carrier IQ (<http://www.carrieriq.com/>).

Carrier IQ is a piece of software installed on millions of mobile phones that logs everything their users do—from what websites they browse, to what their text messages say. The embattled mobile monitoring software maker was secretly installed on some 150 million phones; it has the capacity to log Web usage, and to chronicle where and when and to what numbers calls and text messages were sent and received. The data they capture to their servers from mobiles is vast; the software can monitor app deployment, battery life, device CPU output, and data and cell-site connectivity, among other things.

Carrier IQ was subsequently hit with a lawsuit; Congress demanded answers once it was discovered what mobile data they were capturing. The developer community began uncovering software that had been installed on devices for years. T-Mobile, Sprint, and AT&T have admitted to using it, but none of them had informed consumers how it was used. However, the Wiretap Act under the Electronic Communications Privacy Act of 1986 forbids acquiring the contents of communications without the user's consent.

Carrier IQ may not have felt the gravity of its situation until U.S. Senator Al Franken started to weigh in and demanded answers from the company. Senator Franken pointed out: "The revelation that the locations and other sensitive data of millions of Americans are being secretly recorded and possibly transmitted is deeply troubling. This news underscores the need for Congress to act swiftly to protect the location information and private, sensitive information of consumers. But right now, Carrier IQ has a lot of questions to answer (FTC.gov)."

The privacy notification policy statement should include compliance with various state and federal privacy laws and third-party initiatives, including the Federal Trade Commission Fair Information Practices (<http://www.ftc.gov/reports/privacy3/fairinfo.shtm>), the California Online Privacy Protection Act (COPPA.org), the Children's Online Privacy Protection Act (COPPA) (<http://www.coppa.org/>), Trust Guard™ (Trust-guard.com) privacy guidelines (<http://www.trust-guard.com/>), the CAN-SPAM Act, and the Apple and Google Adsense™ and AdWords™ Privacy Policy requirements. The following provides a starting template for developing a legal privacy notification policy for mobile analysts, marketers, enterprises, developers, and brands:

1. Give users meaningful choices to protect their privacy.
2. Make the collection of personal information transparent.
3. Be a responsible steward of the information being captured.

4. Develop products that reflect strong privacy standards and practices.
5. Use information to provide users with valuable products and services.
6. Provide an option for consumers to opt out from mobile mechanisms used to track and target their devices.
7. Explain why these tracking mechanisms are being used—for enhancing the relevancy and customer service to mobiles.
8. If anonymous techniques and technologies are being used, such as digital fingerprinting, let consumers know so they can make their choice over privacy and relevancy.
9. List what information is being captured by cookies, beacons, digital fingerprinting, and other tracking mechanisms.
10. Total transparency should be the policy to follow in mobile analytics; explain what data is being collected and for what purpose; educate consumers on the benefits of these information collection techniques.
11. Openness is a good general policy for mobile analytics. When a myriad of pitfalls and unexpected dangers may loom, total transparency is the best insurance policy.
12. State the purpose of setting first or third party cookies from ad networks. Mobile marketers should limit the use of third-party cookies, as they dilute proprietary consumer knowledge, which is possibly the most valuable asset companies have.
13. Mobile analytics involves capturing customer events and actions over time and modeling these stored interactions, to determine typical behavior and deviations for the predictions of future revenue growth and consumer relevancy.

3.25 Mobile Big Data

For years, IT departments have stood behind protective firewalls in determining who has access to what. It was a model that mapped well when everything including the user, app, and data was contained in one company. But four megatrends are ripping apart that reality: cloud, social, mobile, and big data. The world looks very different today than it did even five years ago, and so mobile data strategies and solutions must change with it. In the landscape ahead, brands, developers, and marketers need to consider a world where users, apps, and mobile data are flung across the Internet in what seems like a random mosaic. Consumer and enterprise apps live both inside and outside an enterprise; business processes and brands—once exclusively internal—now span cloud, social, and mobile networks…and end-user mobiles.

Within this emerging computing landscape, the old paradigm of control and security—everything safely contained in one place behind a company's physical and virtual walls—completely breaks down. In the new world, enterprises must re-create security and control in a fundamentally different way—one in which distributed systems still appear as one but remain separate. That is where identity factors in.

Identity is core to both freedom and security in today's connected world. Mobile users need the freedom to pass through those old walls, as well as the security to be themselves without having to remember a separate username and password for each resource. Get it right, and enterprises will enable the vision of anywhere, anytime, and any device productivity. Get it wrong, and the cost, complexity, and security implications for IT could very well negate any gains.

In this new world, the objective is no longer centralizing employee information for managing access to enterprise resources. Business cases for identity management, traditionally centered on compliance and security, are now too narrow in scope. And the corporate user directory is but one repository of information that helps identify people and control access privileges. The changing dynamics of corporate computing demand that the current model get a cloud and mobile big data make-over.

Fortunately, the building blocks of identity—namely open identity standards—are reaching a tipping point. Much like the SMTP (Simple Mail Transfer Protocol) for e-mail or TCP/IP (Transmission Control Protocol/Internet Protocol), the networking Internet protocol, these open identity standards promise to unlock the potential of the cloud, mobile, social, and big data for both personal and work-related activities, simultaneously providing users with more convenience and fewer log-ins, more security with stronger authentication, and better control for IT. Mobiles are poised to become the unifying device to synthesize, unlock, and secure both of these personal and professional activities.

The IDC (International Data Corporation; http://www.idc.com/ predicts that in the next three years, more U.S. Internet users will access the cloud through mobiles than through PCs or other tethered devices. The evolution of identity cannot come fast enough. For large enterprises, the shift will be gradual—evolutionary in nature—as the new paradigm must be absorbed as an extension to today's investments in security and control. This will manifest itself in hybrid architectures, where existing control over identity remains within the enterprise, but some use-cases migrate to the cloud for speed, convenience, and cost savings.

For others, companies with fewer legacy dependencies to carry forward, the trend will be more radical as they start in the cloud with the new paradigm and move forward from there. There is a definable transformation under way. It has been slow to build over the course of the past decade; but with key standards now nearly complete, companies are approaching a point of no return. In this new world, people, apps, mobiles, and big data will be distributed, yet seamlessly will be strung together vis-à-vis these identification standards.

3.26 Mobile Data Triangulation Techniques

There is a dizzying array of mobiles in the marketplace, each with a unique set of capabilities and limitations but with two dominant operating systems: Android and Apple. With the onslaught of these mobiles and platforms, advertisers are tasked with creating content that will reach their target audiences—wherever they are and on whichever device they have; this requires creating a mobile analytics strategy and the development of a marketing framework. However, customizing the content and experience for each mobile device type—at their specific location—presents lucrative opportunities for analysts, developers, marketers, and their brands.

Delivering content to these mobiles requires careful up-front planning; consumers expect that whatever device or technology they have, that the content will appear and perform as expected. Mobile device depositories such as the WURFL (Wireless Universal Resource FiLe) or ScientiaMobile (<http://www.scientia-mobile.com/>) databases provide current and frequently updated information on thousands of mobiles, each with hundreds of different capabilities. Once the specific device capabilities and settings have been determined, the most appropriate content can be programmatically selected or generated and served by executing a bit of in-session business logic.

Different mobiles have different screen sizes and support different video formats and protocols such as Flash or HTML5. Some allow third-party cookies by default, which enables behavioral targeting via mobile analytics, while others do not. This is important because other mobile mechanisms must be used, such as digital fingerprinting. Actions to enable click-to-mobile Web, click-to-call, click-to-video, click-to-SMS, click-to-locate, click-to-buy, and click-to-storyboard transition—leading to a subsequent interstitial ad—are all dependent on the capabilities of the device.

Rather than building and maintaining multiple versions of an ad to support different mobiles, a single generic version can often be created and maintained for customized real-time delivery for each device making a request. This is a common design strategy, as it eliminates the need to create and maintain a large number of different content to support different mobiles.

Marketers can also infer demographic information from the mobile device type and customized apps, which can then be used to influence when, where, and how the content or experience is delivered. The data mining of mobile devices extends beyond traditional demographics; as we have seen, there are a host of other mechanisms and techniques for accomplishing this—all of them based on behavioral targeting. For example, research from Google.com shows that nearly half of users who are shown a location-based, time-sensitive, or interest-based ad on a mobile will take some immediate action when compared to other advertising channels.

Today, mobiles have the ability to reach out to a geo-location service such as the W3C Geo-location API (<http://www.w3.org/TR/geolocation-API/>) to determine its current position. The W3C Geo-location API is an effort by the WorldWide Web Consortium (W3C) to standardize an interface to retrieve the geographical location information for a client-side mobile. From this information, the actual location of the device can be detected in real-time, which is typically called "reverse geo-coding" so that the appropriate location-aware ad or content can be served.

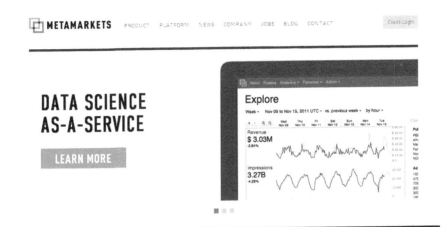

Figure 3.20 Mobile metrics from Metamarkets.

For example, a location-based ad can provide a financial incentive in the form of a coupon, a deal, or a discount for a business or a brand in close proximity to the user's mobile, thereby further enhancing the attractiveness of the offer. Mobile analysts and marketers must build applications that can deliver the appropriate content to maximize the experience for each mobile user, regardless of the device they are on. This may mean streaming video content—in the right format, with the right protocol, and with the appropriate bit rate for specific mobiles.

One option to enterprises, agencies, publishers, and marketers is to outsource this work to such companies as Metamarkets (<http://metamarkets.com/>; Figure 3.20), which offers mobile reporting, benchmarking, and analytics. Metamarkets aggregates billions of mobile transactions to deliver dynamic price data, volume aggregations reports, and analytic media market views. Their Mobile Analytics service basically applies this technology to mobile platforms and sites; they can process, store, and analyze daily mobile events to deliver such metrics as impressions and revenue.

3.27 Mobile Data over Mobile Revenue

A whopping 85 percent of the traffic traversing the four nationwide mobile operators networks is pure data, according to a recent study, showing that the U.S. mobile industry long ago abandoned its voice mantle to become a data-driven juggernaut. The study also found that data accounted for only 39 percent of all mobile data revenue that carriers collected. Operators may be running what are primarily data networks, but they are still getting paid mainly for voice minutes.

A detailed report from ATT.com also shows that overall revenue for operators is increasing, driven by data use, which will grow from a $67 billion market in 2011 to an $80 billion one at the end of 2012. But the average revenue that carriers collect per customer is declining: For every 52¢ of new data fees operators took in last quarter, they lost 96¢ in voice fees. Taken at face value, these numbers paint a rather foreboding picture: The future of mobile operators clearly lies in replacing voice minutes with data megabytes on the earnings sheets, yet voice revenue is declining faster than data revenue is increasing. In addition, operators are using far more network capacity to deliver one dollar's worth of data than they are using to deliver one dollar's worth of voice.

But the numbers also do not paint a precise picture. While 85 percent of traffic may be data, the networks that carry it are orders of magnitude more efficient than they were in the past. For the same infrastructure and spectrum investment, operators can deliver multiple megabits of capacity where they once could offer only dial-up modem speeds or a few dozen phone calls. Those network efficiencies mean operators can make money off data even if it takes over 100 percent of their traffic, as long as they keep upgrading their networks. But the loss of voice revenue is the bigger problem. Voice accounts for so much of their revenue stream, yet so little of their network resources that carriers are obviously using it to pad their profits. If some over-the-top VoIP service were to get widely adopted, it could wipe out a good percentage of customers' bills, forcing operators to build their business models primarily on data.

That does not mean they could not make money on data—it just means that they would make a lot less. A recent study by ATT.com found that only the top 30 percent of mobiles users consume more than 1 GB per month. Yet the data plans that operators sell are designed to give average consumption a wide berth. The smallest mobile data bucket Verizon sells contains 2 GB for $30 per month, while AT&T offers a new pricing structure where customers can buy a 300-MB plan for $20, but the next tier is 3 GB for $30. That means the vast majority of U.S. mobile customers are coming in way under their caps, buying a lot of data they will conceivably never use.

3.28 Triangulation Targeting via Mobile Data

The increasing prominence of social networks as the means to communicate and share information among mobiles and their friends provides many ways to identify, target, and reach new growth revenues and to segment new potential customers via mobile analytics. The trick to data mining mobile devices is not only to track and target their behaviors, but more importantly it is to model them—*you are where you will be*—in a relevant and intelligent manner. In addition, by modeling mobiles' behaviors, marketers and brands can begin to anticipate their future actions and preferences for certain content, products, and services.

To evaluate and improve the performance of mobiles and their dominant operating systems, both Apple and Google collect diagnostic information from randomly selected devices. For example, when an Apple mobile device makes a call, they may determine the device's location at the beginning and end of the call to analyze whether a problem such as dropped calls is occurring on other of their devices in the same area. Google devices also collect location every few seconds and transmit that data to company servers at least several times an hour.

The Apple and Android mobiles are also equipped with GPS chips. A GPS chip attempts to determine a device's location by analyzing how long it takes for satellite signals to reach the device. Through this analysis, the GPS chip can identify the device's latitude and longitude coordinates—such as altitude, speed, and direction of travel, and the current date and time where the device is located. For example, Apple collects GPS information to analyze traffic patterns and density in various areas. The collected GPS information is batched on the device, encrypted, and transmitted to Apple servers over a secured Wi-Fi Internet connection every 12 hours with a random identification number that is generated by the device every 24 hours; his information cannot be associated with a particular individual.

Both Apple and Google got into mobile ad networks early; Google purchased AdMob while Apple launched their iAd. The iAd network offers a dynamic way to incorporate and access advertising within apps; mobiles can receive ads that relate to their interests. For example, a device that purchased a sci-fi movie on Apple iTunes digital store may receive an ad regarding a new movie of the same genre.

Apple also collects information about the mobile's location via its latitude and longitude coordinates—so that when a request is made, for say a sports bar, this information is transmitted securely to the Apple iAd server via a cellular network connection or Wi-Fi Internet connection. The latitude/longitude coordinates are converted immediately by the Apple servers to a five-digit zip code. Apple does not record or store the coordinates—Apple stores only the zip code. Apple then uses the zip code to select a relevant sports bar ad for that mobile.

Apple does not share any interest- or location-based information from their mobiles, including the zip code calculated by the iAd server, with advertisers. Apple retains a record of each ad sent to a particular mobile in a separate iAd database that is accessible only by Apple—to ensure that customers do not receive overly repetitive or duplicate ads and for their own administrative purposes and network regulations. Google will also triangulate the mobile device's location, via GPS and/ or Wi-Fi, in order to provide their Google Maps™ service.

Google and Apple are racing to build massive databases capable of triangulating mobiles based on their locations and interests. These databases will allow them to tap into the $2.9 billion market for an advertising sector expected to rise to $8.3 billion in 2014, according to research firm Gartner (Gartner.com). Marketers and brands can also create their own location- and interest-based Apple and Android apps to leverage the triangulation targeting of millions of mobiles.

3.29 The Mobile Data Challenge

Brands, enterprises, and marketers should be able to give people exactly what they want; that is the benefit of having so much mobile data. But they need to educate consumers about their mobile data—without scaring them. The argument should be the convenience and relevancy they will get in return. Consumers are happy for their data to be out there—in such social networks as Facebook—but they want some value in return. Mobile operators, marketers, and brands have a great opportunity to generate propositions around real-time mobile data. Mobile operators should agree to commercial partnerships with brands and other mobile players to create real-time value propositions for consumers, for they have the power to turn real-time insights into messaging opportunities.

Mobile operators are in a great place to drive end-user propositions; however, there are questions as to what extent these carriers can or are using the data they have access to—mobile data may be sitting in the wrong place. Google and Apple need to consolidate and share their mobile data. There is value locked away in mobile data but, because players are not sharing, the opportunity is being missed. Closed gardens are not in the consumers' best interests. If companies captured data and used it smartly, they could provide real value to consumers, and combined data would work best but mobile players have not recognized the possibilities yet. Operators and other players need to make the consumer experience easy through partnerships. Consumers do not want to be hemmed in—they want freedom and relevancy in their mobile data.

3.30 Triangulating Exchanges of Mobile Data

One of the ways to market to mobiles is via the use of ad networks and exchanges. These networks manage and run ad campaigns for mobile brands. With the number of devices on the rise, the mobile ad market is expanding amid signs of a broader shift in the advertising industry toward mobile.

Mobiles matter; paid-search ads targeted at mobiles provided advertisers with a 37 percent higher click-through rate compared with stationary devices—and at a lower average cost, according to a study by the MarinSoftware (Marinsoftware. com) ad platform. Consumers seem to click on mobile ads at a higher rate for a lower cost to ad men, thus providing a better return on investment (ROI). Marin used Google data for the case study it conducted for its clients; metrics tell the tale of a growingly aggressive mobile ad market.

The following are some key mobile ad practices to better engage and convert consumers via their mobiles. These practices evolved from observed trends and patterns within iOS and Android ad campaigns. The bottom line for mobile ad men is to capture the user's attention and deliver key messages via quick burst calls-to-actions for gaining consumer engagements and conversion:

1. Ad mobile videos should be 15 seconds or less.
2. Calls-to-action must be clear: "Watch This" and "Try This."
3. Teaser and short animations increased engagement by 40 percent.

4. Geo-targeting and offering driving instructions increased engagement by 27 percent.
5. Offer "email this to me," "email a friend," and/or "send me a text later;" users are on-the-go.

Mobile ad network solutions are advancing by leaps and bounds, particularly concentrating on interactive combinations of location- and interest-based apps and personalized recommendation from mobile sites, as well as mobile ad networks and ad agencies. Here is short list of the best ones:

www.mobsmith.com – A turnkey mobile advertising platform.
www.01tribe.com – Italian brand mobile ad agency.
www.12snap.com – German mobile marketing company; it promotes Nokia devices.
www.2ergo.com – Focuses on new media; customers include Fox News.
www.5thfinger.com – Funky mobile marketing agency of Victoria's Secret.
www.acision.com – Offers cross channel advertising via its own platform.
www.addictivemobility.com – Socially targeted mobile agency.
www.ansiblemobile.com – Full-service mobile agency; clients including Intel.
www.buongiorno.com – A division of Buongiorno; offers mobile advertising campaigns.
www.carat.co.uk – Media planning and buying agency owned by the Aegis Group.
www.cellempower.com – Middle eastern mobile marketing provider.
www.fetchmedia.co.uk – International mobile marketing agency whose clients include Polydor.
www.gomeeki.com – Mobile Web agency also offering mobile marketing.
www.grey.com – Full-service agency part of WPP (Wireless Packet Platform) group; clients include Coca-Cola®.
www.groupm.com – GroupM owns the Mindshare mobile agency.
www.iconmobile.com – Design agency from Berlin that has its own service platform.
www.imagineww.com – Serves over three billion mobile impressions in ninety-five markets.
www.insidemob.com – Offers management of entire mobile advertising campaigns.
www.jinny.ie – Offers a campaign engine to brands and operators.
www.marvellousmobile.com – Plans, tracks, and reports on mobile advertising campaigns.
www.mcsaatchimobile.com – Part of Saatchi, a leading full-service mobile ad agency.
www.mindshare.com – A multinational advertising agency.
www.miva.com – A wide-spectrum Web and mobile advertising firm.

www.migcan.com – Offers mobile advertising and marketing.

www.mo2o.com/en – Pioneers of mobile marketing in Spain.

www.moblin.com – Israeli ad company; clients include Microsoft, Doritos®, Nike, and Nestlé.

www.ogilvy.com/o_one – The mobile branch of Ogilvy advertising agency.

www.phonevalley.com – Mobile media planning, interactive services, sites, and apps.

www.plantain-media.com – Seattle-based mobile marketing company.

www.plasticmobile.com – Canadian company with strong portfolio of clients.

www.publicis.com – Multinational mobile marketing group.

www.ringringmedia.com – UK mobile advertising agency.

www.spongegroup.com – Uses any mobile channel available for brand ads.

www.sponsormob.com – International network for mobile ads and apps.

www.textopoly.com – Rich media mobile apps and ads.

www.thehyperfactory.com – Specializes in long-term mobile ad campaigns.

www.useradgents.com – It creates mobile campaigns for brands, apps, and sites.

www.velti.com – Seventy ready-to-use ad campaign templates on its mobile platform.

http://www.vdopia.com – Operates the largest video mobile ad network.

www.welovemobile.co.uk – A London-based mobile ad agency.

www.wpp.com – A multinational marketing company operating in the mobile ad sector.

www.adfonic.com – A self-service mobile advertising marketplace for mobile ad men.

www.aditic.com – A rich media, video ads network.

www.admob.com– Acquired by Google; has been merged into AdSense.

www.admoda.com – An ad network that serves mobile or apps.

www.buzzcity.com - A mobile social networking site offering ad placement.

www.casee.cn – The largest mobile ad exchange in China.

www.decktrade.com – A self-service mobile advertising multi-network agency.

www.digital-advert.com – A French mobile ad network for new media.

www.dsnrmg.com – A results-based mobile ad network.

www.advertising.apple.com – Apple's app and ad network.

www.gigafone.com – A Russian mobile advertising player for aggregation of app data.

www.egsmedia.com – French mobile agency for all types of media campaigns.

www.google.com/ads/mobile – Offers reach in the mobile arena and targeting.

http://m.hands.com.br – A Brazilian mobile advertising network.

www.hipcricket.com – Comprehensive permission-based mobile ad network.

www.iloopmobile.com – Offers ad campaigns via customizable mobile sites.

www.inmobi.com – Indian mobile ad network funded by Softbank.

www.jumptap.com – Search market and mobile ad network.

www.leadbolt.com – Platform helps deliver results for publishers and advertisers.

www.madhouse.cn – MadNetwork is China's largest mobile ad network.

www.madvertise.de – German-based ad network via self-service platform.

www.millennialmedia.com – Rich media mobile ads via its own MBrand network.

www.mobiadz.com – Ad-network with a focus on new media and metrics.

www.mobilefuse.com – Mobile ad network with a reach of 85 million mobiles.

www.mobiletheory.com – Empowers top brands to deliver mobile ad campaigns.

www.mobgold.com – Helps advertisers reach targeted users on mobiles.

www.mobileiq.com – Its Fabric platform serves ads to mobiles.

www.mojiva.com – A self-service mobile ad provider.

www.nexage.com – A mobile ad optimization platform.

www.offerpalmedia.com – Monetizes social networking mobile apps.

www.pixelmags.com – U.S. ad network for Apps with HTML 5 capability.

www.quattrowireless.com – Apple platform-based advertiser.

www.thirdscreenmedia.com – A self-service mobile ad network.

www.tmsfactory.com – Creation, distribution, and management of mobile campaigns.

www.todacell.com – Israeli location-based mobile ad network.

www.webmoblink.com – A self-serve marketplace for mobile advertising.

www.widespace.com – Handles ad publishing, analytics, and payments.

www.ybrantdigital.com – Display, in-game, and Facebook ad network.

www.zestadz.com – Offers WAP and SMS advertising through their network.

3.31 The Future of Mobile Data

The National Science Foundation (NSF) has awarded $10 million to the University of California – Berkeley for the purpose of advancing "big data" research and technologies of mobile data. The grant was part of a larger initiative by the Obama Administration that allocated $200 million around the country to big data technology. The UC Berkeley funds will go toward the university's Algorithms, Machines and People (AMP) Expedition, which is already conducting several projects tackling large datasets. Previously, AMP Expedition was mostly funded by the private sector, but the NSF award roughly doubles its operating budget. As more and more consumer activities go mobile, it is becoming very easy to collect very detailed data, but what is hard is actually making sense of that information. Buried in all this mobile data are the keys to understanding what is going on and how to improve things.

The main issue with big data right now is twofold: The size of data overwhelms current infrastructure, and nowadays data comes from many different sources, making it very difficult to compile and understand. Among the projects going on at Berkeley's AMP Expedition that hope to make sense of all this data is a cancer genoming application. The purpose of that app is to analyze the genetic sequences of numerous types of cancers to help those in the medical field create personalized treatments for people suffering from the disease. There is a flood of genomics data that is coming out that researchers are going to have to be able to deal with. Right now, the growth of that mobile data is far outpacing the rate at which computers become faster.

Another project is Mobile Millennium (<http://traffic.berkeley.edu/>), a traffic-monitoring system tracking Northern California as well as Stockholm, Sweden, using 60 million GPS points every day. Mobile Millennium uses crowd sourcing along with other data sources, such as traffic cameras, to monitor traffic and forecast it. While Mobile Millennium is not the only entity that tracks traffic, it does it from a research perspective—not a commercial one. UrbanSim (<http://www.urbansim.org/Main/WebHome>) is another one of the projects at the AMP Expedition. It takes data from different sources including city records and third-party databases, such as lists of businesses, to put together highly detailed models of cities. Right now, that project is collecting data in San Francisco.

The people behind UrbanSim hope that after the project creates a model for a city, it will offer the ability to experiment with different policy decisions and see what would happen. For example, it could show the potential results of a new public transportation hub, or how the city would be affected if a new pedestrian mall was built. The key to data mining mobile devices lies in the use of AI to model their behaviors; but first in the next chapter we explore how several key players are creating different marketplaces and distinct mobile mobs with different pursuits and objectives.

Chapter 4

Mobile Mobs

4.1 The Mobs

First, some of the numbers: mobile marketing revenues will grow to an astounding $58 billion by 2014 (Gartner.com); there will likely be more mobile Internet users in 2015 than PC users (IDC.com); mobile marketing will grow by 30 percent in five years, to $1,047 billion in 2016 (Ovum.com), mobile traffic will increase 10-fold over the next five years (Ericsson.com); and finally, Cisco (Cisco.com) has come up with a forecast that is downright staggering. According to Cisco, mobile data globally is expected to grow 18-fold by 2016; Cisco (Cisco.com) said in the forecast update to its VisualNetworking Index of traffic alone will grow to 50 times more than it is today by 2016, there will be more mobiles (10 billion) than the number of people on Earth, estimated at 7.3 billion.

So what?

It means that data mining mobiles is BIG business. Mobile analytics is about data mining billions of digital devices in people's pockets or purses that are continuously broadcasting several factors most marketers never had—consumers are *telling* brands and marketers: WHERE, WHAT, WHEN, HOW, and WHEN they want from their products, services, or content. Because mobiles are moving, modeling of their data is about analyzing their behaviors, which are everywhere, ubiquitous, and *intimately detailed*, based on users' personal interests, the locations of their mobiles, and social networks interactions and communications.

The data mining of mobiles is about leveraging AI (artificial intelligence) technologies such as clustering, text mining, and decision tree algorithms to predict what content, products, or services mobiles are likely to want. It all comes down to pattern recognition. It does not matter if it is forensics, fraud detection, Web analytics, or now mobile analytics; the important point is throwing the "hand grenade" as close as you can to the target.

Mobile analytics is about analyzing digital fingerprints, mobile cookies, beacons, and Wi-Fi and GPS triangulation data from mobile sites and apps—and the aggregation of the data and the modeling via AI software—leading to the triangulated marketing of relevant content, products, or services to billions of consumers. The mobile is the consumer; knowing and modeling its hardware and human attributes can be highly profitable—and that is what brands, enterprises, developers, and marketers should strive for. Mobile mobs are distinct marketplaces, with the main ones being Twitter, Apple, Google, Facebook, and Amazon.

Google is aiming to make billions of dollars by sending mobile ads to individuals based on Web searches. Facebook is working on triangulating location mobile ads based on their users' interactions with their friends. Facebook has the advantage of knowing what its users and friends do, while Google knows about what its users are searching for online. Perhaps the real "mover" in this mobile marketplace will be Amazon and its mobiles, which provide cheap and easy access to the largest retailing ecosystem in the planet—content, movies, books, music, storage cloud space—whatever consumers want. Apple and its iTune marketplace is yet another worldwide retailer of digital goods.

The Amazon devices will be a major conduit to one of the fastest growing retail operations the world has ever seen, funneling users into Amazon's meticulously constructed world of content, commerce, and cloud computing. Amazon Prime can drive purchases of toys, toasters, diapers, etc. The company can also exploit their cloud computing initiative—Amazon Web Services®—enabling owners of mobiles to store as many books, songs, movies, or other content on their cloud servers as they like, and at no charge.

4.2 Deal Mobs

Parallel 6 (<http://www.parallel6.com/>), a leading social media and mobile technology company, launched a new mobile app that offers local deals to Deal Mob customers. Deal Mobs' mobile app was created using the Parallel 6 custom-branded Captive Reach platform. Geo-fencing and segmentation algorithms within the platform allow Deal Mobs to ensure the relevancy of an offer by targeting consumers based on specific demographics. Their platform allows Deal Mobs (now part of Parallel 6.com) to target consumers based on their age, gender, and location.

Once consumers receive a daily or weekly deal, they are able to locate, receive, purchase, and share the deal of interest through social networks, including Facebook, Twitter, and LinkedIn. The Captive Reach mobile app extends beyond social sharing to provide a "mobile wallet" for its users. Users can enter credit card information to purchase deals within the mobile app. Once purchased, the user is

Figure 4.1   The Deal Mobs mobile app.

provided with a redeemable code that is then shown to the retailer at the point of purchase. The Promotions Management Panel, which is integrated into the Deal Mobs mobile app (Figure 4.1), serves as a central hub for content, promotions, offers, and incentives, as well as consumer insight and analytics. Deal Mobs can identify target markets for merchants by creating strong relationships and ultimately expanding their reach nationwide.

Forrester Research (Forrester.com) reports that online purchases made on mobiles now account for 20 percent of all purchases and that nearly 60 percent of consumers use them to shop. The analysis of mobile consumer behaviors involves using machine-learning algorithms that can generate (1) clusters of mobile activity using self-organizing maps (SOMs), (2) extracting key concepts from unstructured content using text mining programs, and (3) developing predictive business rules to quantify and monetize mobile behaviors via the use of decision tree software.

Mobile analytics provide "situational intelligence" and the understanding of not only where a person is and what they are doing, but most importantly, what their "intent is"—or what they might be doing in the future by modeling the behavior of the mobiles in their purses and pockets. Analytics has the capability to model and predict what people will do—not only on search and social networking sites, but also with text mining of messages or calls they make anywhere, anytime. Mobile analytics can develop a comprehensive view of how people connect and search via their mobiles; this can generate valuable insight for customized services, products, and content for marketers, carriers, retailers, and enterprises.

Case Study: Urban Airship

As the world's largest and most authoritative online dictionary, Dictionary.com provides a destination where users can access myriad educational and entertaining vocabulary-building tools, including their customized app developed by Urban Airship (<http://urbanairship.com/>). The company's core app has had over 22 million installs across all major mobiles, making Dictionary.com the world's most downloaded mobile dictionary. Dictionary.com continually upgrades its app, adding innovative features such as voice-to-text capabilities and dynamic content.

Dictionary.com Word of the Day is a main feature of the mobile app. Their goal was to offer a word-discovery experience accessible anywhere, anytime, making such popular features as easy to access as possible a top priority. After assessing high usage and positive user comments about the Word of the Day, Dictionary.com decided to offer push notifications to users, enabling them to receive the Word of the Day without any effort on their part. As Dictionary.com continued to add ancillary features and upgrades to its app, the company wanted to keep its focus on what was core to its mobile business. Urban Airship began delivering Word of the Day via push notifications to millions of mobiles on a daily basis.

As a result of leveraging Urban Airship, Dictionary.com has seen an uptick in active user rates and visits per month to their mobile site. Push notifications by Urban Airship have kept Dictionary.com users engaged with the app, reminding them of the useful and informative content that is available within the app. The success of the push notification program has also been a key component to the Dictionary.com overall brand strategy.

Modeling mobile behaviors also involves the analysis of mobile purchases, which are also known as mobile wallets, and it is an alternative payment method. Instead of paying with cash, checks, or credit cards, a consumer can use a mobile to pay for a wide range of services and digital or hard goods such as music, videos, ringtones, online game subscriptions or items, wallpapers, and other digital goods. It can, in addition, include the payment of transportation fares for bus, subway, or train, and even parking meters and other services such as books, magazines, tickets, and other hard goods. There are four primary models for mobile payments that can be analyzed:

1. There is the premium SMS-based (Short Message Service) transactional payments, where the consumer sends a payment request via a text message and a premium charge is applied to their phone bill or online wallet. The merchant involved is informed of the payment success and can then release the paid-for goods.
2. There is also Direct Mobile Billing (DMB), wherein the consumer uses the mobile billing option during checkout at an e-commerce site, such as an online gaming site, to make a payment. After two-factor authentication involving a PIN (Personal Identification Number) and One-Time-Password, the consumer's mobile account is charged for the purchase.
3. There are also mobile Web payments where the consumer uses webpages displayed or additional applications downloaded and installed on the mobile phone to make a payment. It uses the Wireless Application Protocol (WAP) as the underlying technology and thus inherits all the advantages of ease-of-use, high customer satisfaction, and follow-up sales.
4. Finally, there is the contactless Near-Field Communication (NFC) Protocol, which allows for simplified transactions, data exchange, and wireless connections between two devices in close proximity to each other, usually by no more than a few centimeters. It is expected to become a widely used system for making payments by mobiles in the United States. Many devices currently on the market already contain embedded NFC chips that can send encrypted data a short distance ("near field") to a reader located, for instance, next to a retail cash register. Consumers who have their credit card information stored in their NFC mobiles can pay for purchases by waving their device near or tapping them on the reader, rather than using the actual credit card.

Mobile payment has been well adopted in many parts of Europe and Asia. The combined market for all types of mobile payments is expected to reach more than $600 billion globally by 2013, which is double the current figure, while the mobile payment market for goods and services—excluding contactless NFC transactions and money transfers—is expected to exceed $300 billion globally by the same year.

The modeling of these "purchasing behaviors" by mobile analyst enables them to track and predict what consumers will do—from the start of a shopping browsing action, to the price comparison phase, to the eventual purchase of a product or a service. Over the next few years, banks, merchants, start-ups, credit card companies, mobile carriers, as well as PayPal, Amazon, Facebook, Apple, and Google, will battle to deliver "mobile wallets" and control the multitrillion-dollar U.S. payment industry.

Modeling mobile device behaviors can not only involve purchases, but also the sharing of information between multiple devices via social networking where one or more individuals of similar interests or commonalities can converse and connect with one another using their devices. Much like Web-based social networking, mobile social networking occurs in virtual communities. As the number of mobiles overtakes the number of laptops and desktops, the evolving trend for social networking websites such as Facebook will migrate to a purely mobile mode.

Case Study: Foursquare

Foursquare (<https://foursquare.com/>) is a popular rewards app where mobiles can "check in" at shops, airports, retailers, restaurants, etc., for discounts and deals; the following are some case studies:

■ Starbucks: Starbucks extends local-store "mayor" specials to all their stores with a single $1 coupon for any Starbucks mayor at any of their stores.

■ BART: The Bay Area Rapid Transit offered a BART-themed badge that can be unlocked by regular riders of the transit system, which awards $25 promotional tickets each month to riders chosen at random who had logged on to Foursquare check-ins at BART stations.

■ BravoTV: Bravo offered Foursquare user badges and special prizes when viewers visit more than 500 Bravo locations. The locations were picked by Bravo to correspond with selected Bravo shows.

■ MetroNewsCanada: MetroNewsCanada, Canada's number-one free daily, added their location-specific editorial content to their Foursquare app. People who choose to follow Metro via their app will then receive alerts when they are close to a Canadian city or province. For example, someone close to a restaurant that Metro has reviewed would receive a "tip" about that restaurant and have the ability to link to the full Metro restaurant review.

■ The History Channel: The History Channel created tips on Foursquare that share historically significant facts with users when they check into a location of note—for instance, the first building that bought an Otis elevator.

■ Golden Corral: The Foursquare mayor can eat free once per day. Check-ins on other location services gives them a chance to eat free as well, with a chance to win an iPad as part of a larger contest than the daily eat-free special.

■ Harvard University: Harvard University encouraged students to rate campus venues, share tips, and work to earn the Harvard Yard Foursquare badge by checking in to a

certain numbers of locations. They also left tips at campus locations for students and visitors alike to explore.
■ The Today Show: Foursquare users who head to 30 Rock will be able to check in and earn three custom badges: the "Newbie" badge for first-time check-ins, a "Roker" badge for three check-ins, and a "10 to 10" badge for those who check in at the Plaza concert series ten times or more.

In parallel, native mobile social networks have been created, such as Foursquare and Gowalla. Initially, there were two basic types of mobile social networks. The first is where companies partner with wireless phone carriers to distribute their communities via default start pages on mobile browsers such as Juicecaster (http://www.juicecaster.com/>). The second type is where companies and retailers have no such carrier relationships—also known as "off deck"—and rely instead on other native methods to attract users.

Case Study: Gowalla

1. *National Geographic*: Gowalla (which has now ceased operation) expanded their trips feature, including fifteen walking tours to explore destinations such as the Seine in Paris, the Avenue of the Arts in Philadelphia, and San Diego's Balboa Park.
2. Chevrolet: Chevrolet and Gowalla partnered up at SXSW in Austin, Texas, to give users who checked in when they arrived at the airport a free car ride downtown to their hotels.
3. *Washington Post*: The newspaper has created its own adventures with trips designed to help travelers discover attractions and explore Washington, D.C., including all the national museums, the Mall, and other treasures with maps of the Metro.

While mobile Web evolved from proprietary mobile technologies and networks to full mobile access to the Web, the distinction has changed into two major types of mobile social networks: (1) Web-based social networks being extended for mobile access through mobile browsers and mobile apps; and native mobile social networks with a dedicated focus on mobile use such as mobile communication, location-based services, augmented reality—requiring mobiles and technology.

The use of mobiles for offline shopping has increased dramatically over the past few years, and even if consumers are not actually making a purchase with their mobiles, they are often using them to research products and prices while shopping. A recent Motorola survey in the United States found that customers have good reason to compare prices on their mobiles devices: 43 percent of respondents said the mobile device improved their offline shopping experience, while 87 percent of retailers said that customers would be able to find a better deal by using their mobiles.

There are a number of mobile apps and websites that enable in-store shoppers to check and compare product prices, but the eBay (<http://mobile.ebay.com/iphone/ebay>) and Amazon (<http://www.amazon.com/gp/help/customer/display.html?ie=UTF8&nodeId=200557220>) price-comparing apps represent possibly the biggest single threat to offline retailers. Using the barcode scanner on their apps or entering a search term, customers can easily check the products they are looking at in a store or their websites. Because so many people have accounts with Amazon, they already have their payment details saved, such as their credit card information—added to the fact that Amazon is very competitive on price—sales can easily be lost to the retail giant.

Mobiles and cloud computing are converging in the modeling of device behaviors; in only a few years, there will be over one trillion cloud-ready devices. Manufactures of new devices are being manufactured with sensors and the hardware necessary to send an SMS message to a local Wi-Fi for delivery to an address in a mobile cloud account. These types of machine-to-machine (M2M) communications, also known as "telemetry," will prevail in a wide range of market sectors in the near future—in everything from manufacturing, retailing, medicine, transportation, finance; mobile miners can model these M2M behaviors in order to anticipate them—and take proactive counteractions.

Gartner (Gartner.com) predicts that mobile cloud computing will reach a market value of US$9.5 billion by 2014. The availability of cloud computing services in a mobile ecosystem incorporates many elements, including retailing to consumers and end-to-end mobile broadband-enabled services. Cloud computing enables convenient, on-demand network access to a shared pool of configurable computing resources—such as networks, servers, storage, applications, and services—that can be rapidly provisioned and released with minimal management effort or service provider interaction.

Not only will retailing and anonymous advertising anywhere anytime be impacted, but work patterns and habits will also change because of the mobile cloud. Experts surveyed by the Pew Internet Project (Pew.com) think that by 2020, most people who use the Internet will work primarily through cyberspace-based applications on remote servers accessed through networked mobiles. Mobile analytics will enable the modeling of work patterns to improve productivity and improve efficiency.

4.3 Mobile Analytics: What, Where, How, and Why

Mobile analytics is about data mining mobile device behaviors, at the server-to-device level, anonymously and ubiquitously. It is about reacting to consumer needs and desires with precise, personal, and relevant content and offers at the microsecond they occur via their mobiles. The strategy for the mobile miner is to gather and model device data; this requires capturing, analyzing, and acting on moving consumer actions via the analysis of mobile websites and mobile apps. This generally involves the following steps:

1. The monetization of mobile behaviors via analytics of mobile sites and apps
2. The use of mobile cookies, triangulation, and hyperbolic positioning
3. The use of digital fingerprinting for segmentation of mobiles
4. The use of mobile social media WOM (word-of-mouth) and data harvesters
5. The use of AI (artificial intelligence) to predict mobile behaviors

Mobile miners need to map a strategy and set clear measurable objectives. They must be fully aware of the available software, networks, and solution providers in order to execute their tasks and mobile ad campaigns. The mobile environment is a rapid, self-evolving marketing ecosystem in which consumers drive demand, product design, service features, and price structure.

And now to the *what, where, how,* and *why* of data mining mobile devices:

What: the behavior of mobiles. The data miner needs to be aware and responsive to how consumers use their moving devices – where and when they search, share, and shop for specific services and products. The same hold true for content via the multiple channels the mobiles and their apps use to read and research for information.

Data mining of mobile devices is about marketing on the basis of frequency, device type, recent activities, preferences, historical behavior, keywords, and location, all of which are critical in the triangulated modeling via inductive and deductive analytics of mobiles. Consumers demonstrate what is useful and valuable to each of them by their actions in search and social sites via their mobiles; the role of the mobile analytics is to deliver and fulfill, if not anticipate these consumer needs, on any mobile anywhere, at any time.

These days, using analytics tools and software to merely track mobile site visitors is not enough. More and more businesses are turning to tools and methods that not only track site visitors, but also monitor their motivations for staying, leaving, or purchasing. Known as behavioral analytics, this combination of tools and methods focuses on the what, why, and how of mobile behaviors. More and more businesses are investing resources to find out how visitors are using their mobile site. The reasoning behind this is pretty simple: if they know why someone left their site without purchasing, then they can take the necessary steps to fix problems and increase conversions.

Crazy Egg (<http://www.crazyegg.com/>)is a mobile Web application that shows the miner heat maps of where users clicked—whether or not there was an actual link there. This is extremely helpful to find out if there are images or areas of a page that *should* be links and are not. The miner can view results in the form of heatmaps to see popular spots on webpages, or as "confetti" to sort individual referral sources.

Yet another tool, Clicktale (<http://www.clicktale.com/>) also shows clickmaps but takes things up a notch by taking video footage of user visits. Imagine being able to watch how a visitor is navigating via a site's content. The results are eye-opening and can help the analyst make the right improvements.

The easiest sales are made to existing customers. KISSmetrics (<http://www.kiss-metrics.com/>) lets the developer follow actual customers' lifecycles. KISSmetrics data is near-real-time, which means developers can make a change in the morning and start seeing the impact of it by afternoon. They also have a live view so developers can see what is happening at their site while it is happening.

Where: on the Web, social networks, and the wireless worlds where mobiles roam. Increasingly, Apple, Amazon, Microsoft, Google, and Facebook are going mobile, and miners and marketers need to develop strategies for knowing and anticipating how mobile traffic develops and how it evolves in the scheme of their mobile ad campaigns, websites, and apps.

As mobiles become more pervasive, the nature of interactions between users and mobile sites and servers is evolving. Mobile apps are becoming increasingly autonomous and invisible—by placing greater reliance on knowledge of context in which they are running and reducing interactions with users. For mobile miners, this translates to an emphasis on "anonymous advertising apps anytime anywhere" to these moving devices in highly dynamic environments—in which resources, such as network connectivity, data aggregation, and modeling becomes a seamless process—driven by highly precise digital behavioral mobile analytics.

Case Study: Skyhook

The Skyhook Location Engine is used by Priceline in its Android app to augment the native location system in all mobiles. The Hotel Negotiator app lets travelers use their mobiles to quickly find and book last-minute hotel rooms at up-to-the-minute current published prices. Or, for deeper discounts of up to 60 percent, they can bid on hotel rooms using Priceline's Name Your Own Price® hotel service. Hotel rooms can be booked up until 11:00 p.m. ET on the night the room is needed. Skyhook provides that needed level of location accuracy and will allow Priceline to build further enhancements into its Android app.

Hotel Negotiator makes booking especially easy and convenient for travelers. The app can help users locate and book a local, top-rated hotel quickly and easily at the best price.

Skyhook technology enables Priceline's app users to know precisely where they are in a town and where their best hotel options are located. Priceline believes the highly accurate location information provided by Skyhook will significantly enhance the overall experience and satisfaction among Hotel Negotiator app users.

How: for the mobile data miner, this means the use of AI software and Web and wireless mechanisms, networked strategies, and modeling techniques. The modeling of mobile behaviors is the key to executing mobile analytics—coupled with the use of apps to know how and when to market to mobiles.

The modeling of mobile behaviors differs from traditional or Web retailing in that it factors in additional facets involving *location-based* and *interest-based* attributes—along with real-time transactions—and the unique features of mobiles. Mobiles drive consumers to engage in moving transactions involving three inter-related behaviors: (1) the mobile getting product or service information; (2) the mobile providing some type of consumer information, such as its location; and (3) actually making a purchase with the mobile itself. These mobile behaviors are what the data miners must be cognizant of and include as part of their predictive models.

Targeted actions such as scanning, sharing, and comparing engage mobiles, and downloaded apps can directly generate revenue when users select a 'Buy' button. The data miner will want to identify this propensity to make a purchase, as opposed to those users who simply scan products but do not actually make a purchase. Classification software such as decision tree programs can be used to perform this type of segmentation analyses for increased sales, consumer relevancy, and loyalty.

Predictive models can be developed to discover the key features of mobile behaviors that identify those devices most likely to make purchases for growth revenue. Retailers can use these segmentation analyses to discover what mobile behaviors lead to the highest conversion rates. By segmenting users by their behaviors, they can discover what mobile scanned products—and which behaviors—eventually lead to sharing with other devices, and more importantly, which lead to mobile sales. The same holds true in terms of providing relevant content to these mobiles, which may be looking for a particular service such as price comparison features as those provided by several companies, including Amazon and eBay.

Case Study: eBay

EBay Mobile (<http://mobile.ebay.com/>) is the company's multiple OS app that was developed once the firm realized that its next major area of growth was in mobile shopping, and made significant strategic investments in mobile commerce. MobileBreakthru was brought on board to design, develop, and deliver on eBay's mobile strategy. eBay Mobile currently accounts for $650 million in annual revenue, with over 200 percent increase in revenue in the couple of years since its release.

Mobile Breakthru led the redesign and deployment of eBay's mobile websites for twenty-three carrier partners, thirteen countries, and seven languages. They managed remote development and design teams, and developed product requirements that defined key features such as commerce-grade encryption, SMS integration, custom mobile detection, and a new buyer/seller dashboard with notifications.

eBay Mobile is a "killer app" with its price-comparison feature. Given the booming mobile usage and relative convenience, eBay is conducting a major pitch to consumers, encouraging them to increasingly shop through their mobile app. Recent product upgrades included the launch of the RedLaser app (<http://redlaser.com/>), which allows for instant comparison shopping through barcode scanning. eBay also stands to gain in its payments segment, given that PayPal has invested heavily this year to improve its mobile payment capabilities through acquisitions such as that of Zong (<http://www.zong.com/>).

With eBay now integrating PayPal into its RedLaser app, users will be able to shop directly on their app instead of actually paying for it at the counter. This can prove to be a significant threat to the Google foray into the mobile payment space and its Google Wallet.

Why: with increasing numbers of mobiles around the world, users are demanding immediate access to relevant mobile data in this expanding channel. Mobile users want Web-based email, news, weather, games, books, sports, and various forms of entertainment—all of which data mining and triangulation marketing can deliver by segmenting and classifying mobile behaviors.

Mobile miners and marketers need an analytic solution that is built to handle the challenges of an ecosystem where multiple mobiles and operating systems compete for eyeballs. With the rapid rise of mobile social networks, viral channels, payment processing options, and revenue streams, accurate data mining analysis is even more crucial for making sense of the mobile market and gaining insights into user behavior.

Mobile analytics needs to process millions of moving events a day in different locations in order to develop a dynamic moving predictive model of consumer behaviors. This requires the tracking of mobile events with timeline views for viral tracking of different mobiles with multiple operating systems. This may involve event tagging inside apps—which may require custom tagging for each event along different timelines—to segment new versus returning users.

It may also involve the segmenting of user behaviors along different time slices, including total time mobile sessions, time since installation and activation of apps, and the time spent on a mobile site. These trends will help everyone develop a more sophisticated behavioral approach to data mining mobiles.

The mobile miner needs to understand that sometimes pulling insights from mobile analytics data can be akin to digging holes—and lots of them. Insights do not come easily, and part of what makes someone an effective analyst is the intuition and perspective necessary to determine *what* data to look at and *where* to start digging. Mobile miners need to spend more time understanding *why* behavioral patterns occur, instead of trying to find them.

While mobile miners may be able to understand mobile behaviors through mobile analytics data, they still have a long way to go before they can understand *who* those customers are. Certain verticals, such as retail, travel, and finance, are accustomed to understanding their customers over multiple touch points, because their customers hand over their data at the time of their purchases, whether via their app or on their mobile site.

For industries that focus on branding, the future may be in connecting on-site activities and sources of traffic to personas identified by third parties on the Web. These personas, such as Nielsen and their Claritas PRIZM® product (<http://www.nielsen.com/>), can be combined with behavioral analysis from third-party cookies, such as those from DoubleClick (<http://www.google.com/doubleclick/>). The merging of third-party data can be a powerful tool for learning more about what mobiles are coming to a mobile site, and the factors that may be driving their behavior.

The truth is that modern mobile Web analysis tools often fail to capture the value of branding campaigns, while brand marketers do not give enough attention to understanding how on-site activity insights can improve the quality of ad buys. It is essential to initiate the conversation between mobile ad buyers and mobile Web analysts *before* the campaign, in order to reach an agreement of how success will be measured.

4.4 The Social Mobile Mobs

Social mobile analytics can also involve gaining insight into how users are connecting, interacting, and behaving inside a specific application. One such company that provides mobile analytics is Kontagent (http://www.kontagent.com/; Figure 4.2); their platform, called *k*Suite, is designed to provide social data pattern visualization and analysis for mobile applications. Another is Xtract (<http://www.xtract.com/about-us>), a company that combines social, behavioral, and mobile

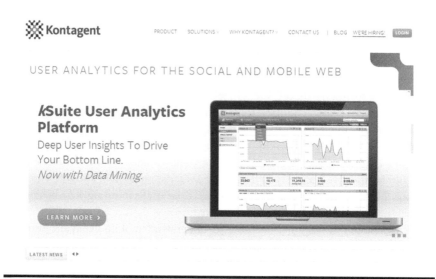

Figure 4.2   Kontagent also offers data mining.

analytics to attract and acquire consumers. It offers what it calls "Social Links," either as a licensed software product or as a hosted service. Their Social Links technology analyzes behaviors, interactions, and demographic data from the user's subscriber network to identify the best targets for effective and personalized marketing campaigns.

Another mobile analytics firm, Flurry (http://www.flurry.com/; Figure 4.3), expects that users will download 25 billion mobile apps this year, a 300 percent increase from a year ago. The company offers software agents and developer tools to enable any mobile app to be tracked by its proprietary software-as-a-service (SaaS) analytics engine. Its AppCircle (<http://www.flurry.com/appCircle-a.html>) is a service that recommends apps to users based on their current app selections. Companies pay for the integration with Flurry apps to get their wares recommended as Flurry locates likely prospects. No personal data is ever sold or revealed, and Flurry offers this SaaS from its own facilities. Kontagent, Flurry, and other SaaS firms are all starting to support these social mobile market analytics. Another start-up is Localytics (<http://www.localytics.com/>), which is also performing mobile cloud-based analytics.

Mobile analytics opens up the possibilities for a market-transact-manage-empower life cycle in which GPS chips, cell-tower triangulation, and knowledge of users' IP and network addresses introduces the door to location-aware services that can respond to, or even anticipate, a mobile user's needs. Mobile miners can provide visibility into what mobiles do—not only on social networking sites such as Facebook or Twitter, but also with text messages or calls from their mobiles. All are important in gaining a comprehensive view of how people connect, as all represent

Figure 4.3 App analytics from Flurry.

social interactions involving mobiles and messaging that generate valuable insight for customized services and content as well as targeted product and services offerings, based on the data mining of mobile devices.

Already, mobiles are redefining the retail shopping experience and will be a $5 billion market per year by 2015, according to a research study by analyst firm IHLGroup (IHLGroup.com). The advent of mobiles is a Gutenberg moment that is revolutionizing many aspects of the shopping experience. Lowe's® (<http://www. lowes.com/>) is arming its employees with mobiles to check inventories and make suggestions to consumers. PacificSunwear (<http://www.pacsun.com/>) uses mobiles to create outfits for customers, while Brookstone (<http://www.brookstone.com/>) is using devices to demonstrate products that can be controlled using their mobile app.

Another area where consumers will see more mobiles is in table-service restaurants. Customers will be able to place their own orders on mobiles at the table as well as access games and other entertainment. More important to the mobile miner, patterns will be discovered and leveraged to increase cross- and up-sales to retailers and restaurants. This will lead to more efficient turnover—restaurants that have deployed mobiles have seen a 25 percent increase in the number of times they can turn the table, thus serving more customers during each shift.

According to IHLGroup.com by 2015, over 2.7 million mobiles a year will be shipped for use in North American retail and hospitality market sectors, an increase of 450 percent. For years, retailers have been looking for technology to both provide associates with more data on the products they are offering, and recognize and reward customers as they come through that door. Mobiles are making this possible from these new moving mobs.

We are at the tipping point of a mobile revolution in which these moving mobs are exploding globally. A radical transformation is taking place in which carriers, enterprises, marketers, and data miners are making strategic plans. They are drawing up their action checklists and procedures on how to model mobile behaviors: What they have to do—not only at the front end with the mobile and cloud environment, but also on how to aggregate mobile behavior data for data mining mobile devices.

The mobile mobs are revealing some general patterns of activity about consumers and their propensity to purchase certain products and services. Grocery, clothing, and entertainment purchasing continue to lead as the categories consumers are deciding to purchase via their mobiles. Larger proportions of consumers are increasingly making decisions about buying travel, financial services, or automobiles. A large proportion of consumers indicate that they would browse a website as a result of seeing a mobile ad. More consumers have indicated that they would redeem or download a coupon as a result of seeing an ad on their mobile.

Mobile marketing is having increasing influence and a direct impact on consumer behavior worldwide. This shows that mobile marketing has made tremendous strides, and consumer activity and interests are clear indications of the larger trends happening in this mobile market sector. Given the explosive growth of mobiles in retailing, Apple, Facebook, Google, Twitter, and Amazon are all rolling out their own "marketplaces," with each providing unique features to their mobile mobs.

4.5 The Apple Mobile Mob

Starting with its musical device iPod, the Apple marketplace started by offering songs for 99¢ over its iTunes platform, gradually expanding to include movies and TV shows. This would become the core engine for the creation of its App Store, which became home to thousands of innovative programs for mobiles. By 2010, Apple had sold 130 million iPhones, accounting for almost half of their revenue; that same year, the iPad was introduced, quickly selling 30 million tablets with its multitouch technology, derived from a company called FingerWorks® that Apple acquired in 2005.

More importantly, the Apple marketplace maintains databases that are continuously updated with location information about their mobiles. Clearly, this enables Apple to target its devices with ads based on when and where they are located in their moving marketplace; it also enables Apple (Apple.com) to know what content consumers desire based on the purchases they make at their iTunes portal (Figure 4.4). Mobiles are redefining the retail shopping experience and will be a $5 billion market per year by 2015, according to a research study by analyst firm IHL Group. According to that study—Mobility: A Gutenberg Moment for Retail, A Threat to POS—the release of the Apple and the Amazon mobiles has created

Figure 4.4   Apple® iTune® store.

price points and form factors for mobiles that are finally allowing retailers to arm their associates with tools that will transform the consumer's in-store experience.

The desire for the "Apple Store experience" among retailers are driving them to look at mobile technology to both provide sales associates with more data on the products they are offering, and recognize and reward customers as they come through that door. Mobiles are making this possible. More importantly for the Apple mob, their iOS provides a unified experience across all of their mobiles.

Nielsen (Nielsen.com) recently reported some new statistics on the iPhone, pointing out that, as of 2012, about one in three mobiles globally are using Apple mobiles. Nielsen also said that the Apple mob is also a top destination on the Web, with 72 million unique U.S. visitors to their site during April 2012, and "has been among the top 10 Web brands overall." Visitors to the Apple website spent, on average, nearly an hour during the month. The analytical firm also said that Apple users download fifty apps on average and that Facebook is the most used app, followed closely by Maps. A third of the Apple mob downloaded a paid app in the past 30 days.

The success of Apple is due to broader distribution—adding Sprint and other carriers, having lower-priced models—and additional distribution partners such as Walmart, Amazon, and Best Buy, and the fact that Apple mobiles cost about the same as competing products. And, based on recent news, it could get even better for the Apple mob, with Virgin Mobile (Virgin Mobile.com) now selling Apple mobiles on a prepaid plan with no contract. The Apple iPhone itself will set users back $649, but the total cost of ownership over a two-year period, according to PCWorld, will be $1,369, which is $300 less than AT&T and more than $500 less than Verizon and Sprint.

One advantage of the Apple mobiles over their Android competitors is a simpler number of choices compared to the highly fragmented Android market. While having a lot of choice and competition in the Android market has some advantages, having a single screen size for Apple mobiles and an ecosystem of accessories that mostly work across the line, has its advantages, as does the enormous Apple app ecosystem. Apple has consistently ranked at the top of J.D. Power and Associates (JDPowers.com) rankings of mobile manufacturers, achieving a score of 839 on a 1,000-point scale, while HTC came in second with a score of 798.

Apple is also assisting mobile app developers in tracking who uses their software. It is the company's latest attempt to balance developers' appetite for targeting data with consumers' unease over how that data is used. The Apple tool aims to better protect user privacy than existing approaches. It comes after Apple rattled the mobile industry by saying it would stop allowing app makers to use a unique identifier embedded in their mobiles to track users across different apps. Many mobile companies rely on what is called the unique device identifier, or UDID, to serve ads and gather data (such as location and preferences) as people use and move between apps. But some privacy advocates argued that the string of numbers, which are anonymous, could be coupled with enough data to identify individuals.

Apple has sent developers flurrying to find tracking alternatives since it began rejecting some apps that take advantage of your mobile's UDID to track user behavior. But now it seems that Apple is gearing up to offer a solution of its own. The solution is that the iPhone maker is developing a new app tracking tool that would provide developers with useful information, while protecting user privacy more than the UDID approach.

It made sense for developers to rely on UDIDs initially; because unlike cookies, which are used on Web browsers to track user activity, Apple's device IDs are permanent, and users cannot block their transmission to third parties. Developers need tracking capabilities to see how users interact with their apps, and mobile ad firms use tracking to target users with relevant ads. But with privacy concerns on the rise, Apple is being forced to come up with a more secure solution. After all, a plethora of other apps got in hot water over accessing the users' address book data, among other personal information. The option for the Apple mob is the use of an opt-in tracking + toggle switch.

The biggest problem with UDIDs for user tracking is that consumers do not actually know their behavior is being tracked. That is great for developers, because it means they can get useful data on just about everyone, but it leaves most consumers in the dark. Apple could fix a majority of its privacy issues by simply making consumers aware that their behavior could be tracked, having it turned off by default, and offering a simple way to toggle the tracking on and off.

In addition, Apple has begun enforcing stricter developer rules. By taking control of the app tracking process, Apple will be able to better keep developers in check. Apple had made it clear to developers that they could lose tracking privileges if they abuse the tool. Apple could still just reject apps outright, but a more granular tool could allow Apple to keep money-making apps available sans-tracking. Per-app tracking—just like how iOS 5 handles notifications on a per-app basis—Apple now offers the ability to tweak tracking by individual app.

The effect is that Apple galvanizes the app developers who make an OS feel vibrant. So far, the Apple mob has resulted in their users downloading more paid-for apps rather than free ones for their mobiles. Apple iPad devices dominated the tablet Web traffic category, accounting for 97.2 percent of total tablet Web traffic. iPad devices also accounted for more Web traffic than iPhone devices in the United States, with 46.8 percent compared to 42.6 percent. Apple's iOS platform also drove more traffic than Android, sporting 43.1 percent of mobile Web traffic compared to 34.1 percent by the Google mobile Android OS platform.

Apple recently dropped Google as the maps provider for their Apple mobiles. Google has provided the Apple mobile mapping system since the release of the first iPhone, but the two companies are no longer friends. Apple has purchased three mapping start-ups, and it has been combining their technology into a maps application. Is Google worried that the Apple defection will substantially reduce its user base of the Apple mob, and, consequently, the advertising revenue it gains through maps? The search company will be losing its place as the online mapping leader, a position that has long been one of its competitive advantages and a good advertising revenue stream.

However, Google recently unveiled a few upcoming features, including a system that creates beautiful 3-D images of all the buildings in the world's major metropolises, as well as a way to access Google Maps when the mobile user is not online. Even so, losing its spot on the Apple mobile home screen is bad news for Google, and it is a move that will be difficult for the search company to overcome. To Apple, mapping will always be a side deal, something it does as a way to improve the iPhone.

For Google, mapping is a critical aspect of its effort to organize the world's information. It is also essential for the Google search engine to be able to deal with a massive amount of queries for physical locations; that is why, when it comes to maps, Google does not mess around. The firm has invested hundreds of millions of dollars in building online cartographic dominance.

Not only does it have a fleet of cars that are constantly snapping photos of streets around the planet, it also has a fleet of planes, and it has also mounted its cameras on boats, bikes, and snowmobiles. Now the company's engineers have created a fully portable camera-laden backpack called the Street View Trekker™; it can be used to photograph places that are not accessible to vehicles, such as the inside of the Grand Canyon, for example.

The new Foogle 3-D imaging capabilities are also impressive. By photographing cities from multiple airplanes on different flight paths, the company can combine and extract 3-D details from flat photographs. The process is called "stereophotogrammetry" and provides results in overhead views of cities that resemble what users would see if they were flying above in a helicopter. Compared to the flat view users now see in Google Maps, the 3-D images look spectacular.

The trouble for Google is that all of its tech advances may be for naught if it is cut off from Apple mobiles, which now make up one-half of the firm's mobile mapping audience. If Apple does drop Google as the default maps app, the search company might need to build its own separate maps app for the Apple mob of mobiles. And even if Apple allows the Google app in its App Store, the company will still have a hard time getting users to choose its program over the Apple native system. Google could also try to convince people to use its browser-based maps system and get them to visit maps.google.com from their Apple mobiles. When they want a map, most people will do the obvious thing and tap Apple's maps app.

The other option for Google is to keep investing more and more into its maps division in the hope that, over time, customers will stick with it because it is the better option. That has worked well for Google in the search business. After all, even if the Apple system does not offer the same 3-D imagery that Google has, and if it cannot provide pictures of the inside of the Grand Canyon, will that really matter to most users? Most people consult a maps app with extremely simple queries: Where am I now? Where am I going? Sure, it will be very nice to fly over a 3-D rendering of cities, but if users are in a hurry and on-the-go, they will not care about that stuff. If Apple manages to make an app that does just the basics; that might be more than good enough.

One of the most startling mobile statistics is the disproportionate usage rate of Apple mob mobiles, which accounts for 97 percent of all tablet traffic in the United States. In addition, analysts at IDC (IDC.com) recently estimated that Apple has sold 75 percent of all tablets. Although the Google Android operating system software has pulled ahead in the mobile race—and more Android mobiles are in use than Apple devices—Apple is still the strong leader in mobile traffic usage. ComScore (<http://www.comscore.com/>) has said that iOS mobiles, which include the iPhone, iPad, and iPod touch®, account for 43 percent of all connected mobiles online. This is an important statistic for the dominance of the Apple mob.

Important to the Apple mob is that the iOS 6 mobile operating system does not support the older Apple mobiles. The Apple software update for its mobiles does not support the original iPad, the original tablet computer of Apple, and also the 3rd-gen iPod touch. The iOS will force the owners of unsupported Apple devices to switch to a new device, like the iPad 2 and the iPad 3 with retina, and the latest version of the iPod touch. The iOS 6 introduced new features that will focus on improved syncing, cloud services, unified store for movies, music, and applications. The iOS 6 can be upgraded to support widgets for the iPhone and iPad. Application widgets support is one of the popular features of Apple's biggest rival, Android.

More astounding is the amount of time people spend online on their iOS devices compared to competitors. ComScore said 56 percent of mobile usage on the Web originates from iOS devices, although Google has activated more Android devices than Apple. This is compared to 32 percent of active mobile traffic originating from Android mobiles. When comparing these numbers, this means people who own an Apple iOS mobile device surf and shop on the Web for considerably longer periods of time than those on Android mobiles. Research In Motion, the maker of the BlackBerry and once the leading mobile device seller, accounted for only 5 percent of mobile surfing traffic.

The Apple iOS mobile operating system, which runs its popular iPhone and iPad, is getting even more user friendly. For example, Siri (<http://www.apple.com/ios/siri/>), the voice-activated personal assistant available on the iPhone 4S, seems to have gotten smarter. It is able to reel off the day's sports scores and statistics during a recent demonstration. Siri will also now launch apps simply by telling it to "Play Angry Birds" or whatever app the user wants it to launch. Siri will be included in the iPad; in addition, the iOS software, also has a partnership with major vehicle manufacturers (BMW, GM, Audi, Toyota, Honda, just to name a few) who will integrate iOS into select cars. A button on the steering wheel allows drivers to access many of the functions on their Apple mobiles hands-free for the Apple mob.

From iOS 6 forward, the Apple mob will have Facebook integration on their mobiles; it will also integrate with the App Store; and FaceTime®, the company's video-chat feature, will be available over cellular networks in iOS 6—not just Wi-Fi. Safari, the Apple mobile browser, is also getting some improvements, with a 3-D traffic map program and real-time information with turn-by-turn navigation, and integrated with Siri.

The App Store has 400 million accounts worldwide, and 650,000 apps, with 225,000 of those specifically designed for the iPad. Apple allows users to post to Facebook from different apps, similar to the level that Twitter integration currently has now. Users will see notifications from Facebook in the "Notification Center," and Facebook events and birthdays will appear in the "Calendar" app. Third-party apps can now be launched with a command to Siri; for example, "Play Temple Run" opens the app. Users can now also tweet from Siri, a formerly noticeable hole in the Twitter integration throughout the rest of iOS.

Siri has become more knowledgeable about restaurants and theaters, with restaurant search results sorted by Yelp (<http://www.yelp.com/miami>) rating, and tapping on them takes the user into the Yelp app. OpenTable® (<http://www.opentable.com/>), a reservation booking app, is also integrated into the restaurant results. For movies, Siri can bring up artwork and the slate of movies playing at a nearby theater, along with information culled from Rotten Tomatoes (<http://www.rottentomatoes.com/mobile/>). Siri can also respond to questions about sports, including queries on standings and player stats.

Apple uses "pixel tags" at its iTunes site. These tags, also known as "beacons" or "bugs," are tiny graphic images used to determine what parts of the site a visitor navigates to; they are silent tracking tags used to measure the activities and behaviors of online visitors at their Apple site. In 2010, Apple launched their iAd mobile advertising network for their devices running iOS4. The iAd network offers a dynamic way to incorporate and access advertising within its apps. Mobiles can receive advertising that relates to their interests, also known as interest-based marketing, or their location, via location-based marketing.

This is how pixel tags work. If a mobile purchases an action movie on iTunes, the user may receive an ad regarding a new movie of the same genre, while a device searching for an Italian restaurant may receive an ad for such eateries once it enters that proximity. Apple has the advantage of knowing when and where products, services, apps, movies, and other content its users have purchased in the past; using mobile analytics, Apple can develop models for predicting future consumer behaviors and preferences.

The same holds true for the thousands of Apple app developers such as Pandora, which upon registration for their apps, asks for the user's age and gender, thus aggregating important consumer anonymous information. Apple and the app developer may in addition pick up mobile device features such as their operating system and hardware version. The point is that all these data points are subject to segmentation and targeting via data mining mobile devices—at an anonymous and highly precise server-to-device level.

The Google Android operating system outsells Apple mobiles more than two to one. And yet, even as the Google system has gobbled up market share, Apple has held onto one critical advantage: the loyalty of mobile app developers. Many developers have continued to make apps first, and sometimes only, for iPhones. They find it easier to create software for Apple mobiles than for ones running Android, or it may be more lucrative. Their allegiance to Apple has helped make their mobiles the powerhouses for their mob.

Android may have a lead in how many handsets it ships, but it does not have a lead in how much money app developers are making from it. Apps are among the strongest weapons Apple and Google have for marketing their mobile technologies to consumers. Rival technologies plagued by a scarcity of apps, including the Research in Motion BlackBerry and Microsoft Windows Phone, are finding it difficult to persuade developers to invest in them. The continued influence of Apple among mobile app developers flies in the face of predictions that the company would steadily lose clout as Android mobiles flooded the market, presenting developers with a much bigger target audience.

In the mobile market, there is no doubt Apple's share has been overshadowed by Google, which makes Android freely available to any hardware maker that wants to build a mobile with it. Although the first Android handsets did not appear until about a year after the iPhone was released, widespread support from handset manufacturers, especially Samsung, and wireless carriers helped propel Android to 59 percent of the mobile market; compared with 23 percent for the iPhone, according to estimates from IDC (IDC.com), a research firm.

Developers say it is easier, and therefore less costly, to develop apps for Apple than for Android phones, in part because there are far more models of Android phones in use, with different screen sizes, processors, and other technologies. The variations in hardware and software are not insurmountable obstacles, developers say, but performing the testing to ensure that apps run properly on most Android phones adds time and cost. Flurry (Flurry.com), a mobile analytics firm, estimated that for every $1 a developer brought in for an application on the Apple iOS, they could expect to take in about 24 cents on Android.

Furthermore, Apple has been far more effective in getting iPhone users to update their mobiles with the latest version of iOS than Google and its partners have with Android. This makes it easier for iPhone developers to write their apps because there is less variation in the underlying software on their devices. One other advantage Apple has among developers is the iPad, which has thus far maintained its dominance in the tablet category, despite challenges by an assortment of Android tablets. Because the iPhone and iPad use the Apple iOS, it is relatively easy for developers to adapt their software to run on Apple tablets, thus significantly expanding the audience of potential customers beyond the iPhone. One occasional source of discontent among Apple developers is the greater control that the company exerts over its App Store, for which it takes a more hands-on approach to approving software for distribution than Google does.

Apple, Google, and Microsoft are expected to be the dominant providers of mapping services for indoor spaces. IMS Research (IMSresearch.com) expects that the mobile giants, already tussling over city views, will next set their sights on indoor spaces such as airports and shopping malls. By 2016, IMS expects nearly 120,000 indoor venue maps to be available to consumers and the Apple mobile mob. Apple acquired Placebase, Poly9, and C3 Technologies for mapping purposes—indoor mapping—representing the next area where Apple and Google will compete with each other.

Microsoft Bing Maps™ began including mapping information for malls and airports, and Google Maps added the same, with the release of Google Maps 6.0 for Android; the latter directory is brought to the palms of users hands via their mobiles, helping them determine where they are, what floor they are on, and where to go indoors. Micello (<http://www.micello.com/>), aisle411®, and PointeInside (<http://www.pointinside.com/>) all have "significant" indoor map databases, with Micello being the market leader by far (Figure 4.5). With advertising revenue generation potential offered through indoor positioning and the impending improvements to indoor location technology on the horizon, these start-ups are ripe for acquisition.

Passbook®, an app in the Apple mobile operating system, gathers loyalty cards, boarding passes, event tickets, and coupons into a single virtual home for the Apple mob. Apple offers it as a way to make shopping, traveling, and seeing movies more efficient. Passbook represents a potential shift in the way the Apple mob pays for goods. Apple allows users to store payment-card information inside Passbook, marking the company's entry into a market that has grown crowded even as it has yet to crack the mainstream.

Apple already has 400 million active credit cards on file in its iTunes store. At the moment, those cards are used primarily to buy digital music, movies, and books. But with Passbook, those same accounts can let consumers buy physical

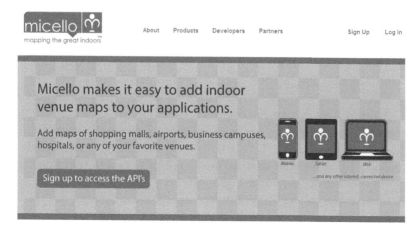

Access over 5,000 indoor venue maps of places throughout the world

Figure 4.5 Indoor mapping for mobiles.

goods from any retailer that accepts credit cards. Passbook is the first step in Apple providing a digital wallet.

The idea of a digital wallet has been more popular in theory than in practice. The same could once have been said about digital music, or tablet-style computers; other markets where early entrants gained little traction have come to see Apple find success by making simple what had once been complicated. Passbook is aware of the phone's location, so when a user who buys a ticket online approaches a movie theater, a notification pops up with a link to the ticket ready to be scanned by an attendant.

Similarly, users who frequent a particular Starbucks® could see their loyalty cards pop up on the screen as soon as they approach the coffee shop. With Passbook, Apple is positioned to really take mobile payments to a new level. Retail data shows that consumers are already growing in their willingness to shop and spend on mobile; it removes some of the previous barriers to spending on their devices. Most companies have relied on NFC technology, which uses a chip inside mobiles to communicate with dedicated merchant hardware.

As previously mentioned, carriers such as AT&T and Verizon, financial services companies such as Visa and American Express, and start-ups such as Square are building digital-wallet systems of their own. If and when Apple decides to enable payments on the iPhone, it will have plenty of competitors. But even if credit cards never make it into Passbook, the app could still give Apple plenty of valuable data about where the Apple mob shop, how often they frequent a given establishment, and what kinds of offers move them to make purchases. With that mobile data, a range of new business opportunities open up for Apple because commerce is not just about payments. Commerce is a chain, from the very beginning of thinking about something a consumer may want to buy to choosing a particular brand, to deciding where to purchase it, to getting after-care. It is a rich chain, and with Passbook, the company is in a position to contribute to or influence its Apple mob.

4.6 The Facebook Mobile Mob

Google, Facebook, and Twitter all now have new platforms they call "Places." The idea of Facebook Places™ is to allow mobiles to share their physical location online with other friends' mobiles. This paves the way for Facebook to become a player in the growing mobile business of supplying local information and targeted advertising based on the location of mobiles and their friends' devices.

While Facebook (Facebook.com) and its 900 million users are often credited for having fundamentally changed the way people view and use the Internet, it is also well known that a large percentage of the site's traffic comes from mobile devices (approximately 33 percent), and understanding mobile is critical for any business these days. In the United States, people are spending about 142 minutes a day on their mobiles, compared to 135 minutes for TV and 96 minutes on PCs, making mobile the primary media consumption channel in the United States alone.

Yet mobile ad spend has not increased at a proportionate rate. This is because mobile marketing requires a completely new way of thinking, a new set of rules and standards, and a fundamentally different strategy. Facebook might have a leg up in its approach to mobile. Facebook has acknowledged something about mobile that no other major player has been prepared to recognize to date: that the growth in impressions available to marketers as mobile penetration grows does not equate to an opportunity to monetize the channel by bombarding people with advertising.

Mobile is personal: The marketing research firm Upstream recently released research (<http://www.wirelessweek.com/News/2012/05/Facebook-future-dependent-on-ipo/>) that showed 72 percent of mobile users in the U.S. find advertisements on their mobiles intrusive. What is more, only 15 percent of Americans who have used their browser on their mobiles have ever even clicked on a mobile ad. To get consumers to respond over mobile, a brand needs to speak less and employ the right social network technologies and formats that achieve true relevance and impact via the user's friends and not brands.

One way to attract visitors is by providing them with mobile apps and that is what Facebook is doing with its App Center, a hub that lets them access apps from Facebook and related third parties. The center started with about 600 apps, including favorites like DrawSomething, Pinterest, and Nike+ GPS. Facebook is taking a creative approach to the apps. In addition to the popular ones, it is suggesting newer apps users might want to try, such as Jetpack Joyride, Ghosts of Mistwood, and Ghost Recon Commander. It is listing only "high-quality" apps to the Facebook mob, a designation based on feedback from people who actually use them.

This being Facebook, there is a social element for users of the App Center. It provides personalized recommendations, letting visitors browse the apps their friends use. There is also a mobile angle, a reflection of Facebook's determination to build up resources in that area. The App Center includes apps for both iOS and Android. There is also a feature that lets the Facebook mob send apps discovered in the App Center to their mobile device. For the Facebook mob, the appeal of the App Center is obvious. A filtered list of apps not restricted to any one platform can only further the goal of Facebook to make its site as sticky as possible.

Rather than manually searching or hearing about new apps from friends, the Facebook mob can now visit the App Center from their desktop computer or mobile the Facebook App Center to see categorized listings of all approved apps. To that end, it has made the App Center as easy-to-search as possible, perhaps as a lure to the Facebook mob who is tired of the cluttered Android Play interface or weary of the seemingly endless selection populating the Apple App Store.

The apps are accompanied by screenshots and a detailed description—including what information the app will require—so users can decide whether they really want to download it. In another nod to privacy, the App Center lets users choose who can see their activity on Facebook. In general, Facebook is offering the same protections of user privacy that it does on its main site, but with the added convenience of the process being integrated directly into the App Center via each app's detail page. Users must grant appropriate permissions to all Facebook apps before using them.

The App Center has a fan base in addition to the Facebook mob, but also the developers that want to target them. The appeal of the mammoth Facebook user base cannot be ignored, and this could be a sweet thing for developers if it helps them gain traction or get the attention of some of the 900 million members of the Facebook user mob. The App Center is a welcome validation that Facebook is still cultivating third-party developers. There was a fear in some quarters when Timeline™ launched that apps would be relegated to Facebook backwaters. Instead, the site has gone on to introduce a number of new tools to help third parties better monitor their popularity on the site, including charting tools that let developers see trends in installations, retention, and churn.

Facebook is stepping up that process with the App Center and is making aggregate user feedback available for brands within App Insights so that app developers can spot trends in negative feedback to improve apps, fix bugs, and continually improve their product in a very social context as expected from Facebook. In addition, there is no preferential treatment given to paid apps versus free ones—it is about how well-liked an app is among the Facebook mob; by making it easier to find apps, more types can become successful.

Facebook is focused on getting the right apps in front of the right people so that developers can spend more time building great apps and less time worrying about distribution. This is important because the use of social networks for marketing continues to grow—increasingly on mobiles. The percentage of company respondents who say they use Facebook for marketing now stands at 84 percent, up from 73 percent last year. For example, Facebook Places (Facebook.com) relies on recommendations from friends to discover relevant content or available products and services. The geographic location of a mobile is critical; Facebook Places lets mobiles share their current location by "checking in." Tap *Places,* and an app in the mobile device can see recent check-ins as well as friends' mobile check-ins; the user can find out more details about the places his or her friends are checking into—map location, description, directions, comments, and other check-ins.

Once the mobile taps the "Check In" button, it will be presented with a list of nearby locations where other "friends'" mobiles have checked in. The user can tap the one he or she wants to check into, and can add comments on what it is doing there or add Facebook friends. While Places is mostly meant for mobiles, it does not depend on GPS triangulation to find the location—the user needs to check in—while anyone can add a Facebook Place, business owners can turn the listing in Places into a proper Facebook Page, with Likes and a Wall and other social media marketing discounts, special deals, and offers.

Facebook Places provides businesses with a platform for marketing and promotions; it also provides them an opportunity to build customer loyalty. Like current social location-based check-in services, Facebook Places lets people share where they are, see which friends are in the local area, and discover new places by following where others from their social network have checked in. The Places app allows friends' mobiles to meet, connect, and share.

For a business to claim their "Place," they need to search for their business name on Facebook via the normal Search bar. Then click on their Places page, at the bottom left side of their Place, where a link will prompt them "Is this your business?" Click on that link and they will be directed to a claiming flow. Facebook will then ask them to confirm ownership of the business through a phone verification process, and they may be asked for some documented evidence. If the claim is confirmed, the business will be able to administer their Place on Facebook.

Facebook has started to roll out a mobile payments service in an attempt to offer a faster and easier method for online payments. The world's largest social networking site introduced the system—a "low friction" mobile payments system brings in a two-step payments system to replace the previous seven-step process. It offers a simpler way for users to purchase Facebook Credits online, a perfect method to buy a new shed in Farmville when needed.

Purchases will simply add on to monthly phone bills, and Facebook says it is available on all mobile networks. The payment flow is simple: users who want to pay for virtual or digital goods in a mobile Web app open the payment dialog and confirm their purchase. Facebook mobile payments service will be easy to integrate for game developers, and mobile Web developers who want to add the system on their sites can use it via the Facebook payments API.

More than 72 million Americans accessed social networking sites such as Facebook (Facebook.com, 2012) or blogs via their mobiles, a figure that represents a 37 percent jump from the same time the previous year, according to data compiled by comScore, a digital marketing research company. The bottom line is that social networking by way of mobiles is on the up-and-up; comScore (comScore.com) estimates that nearly one-third of all U.S. mobile users are now accessing social media services, and that close to 40 million Americans are doing so on an almost daily basis.

However, Facebook will become something of a has-been within 5 to 8 years, according to one analyst's prediction. According to the prediction, the social network will disappear in the way that Yahoo! has disappeared. Yahoo! is still making money—it is still profitable, and it still has 13,000 employees working for it—but it is 10 percent of the value that it was at the height of 2000. For all intents and purposes, it has disappeared. The prediction is based on the fact that there have been three generations of Internet companies. Yahoo, a Web portal, is a great example of an online pioneer. Facebook then swept in as the second generation with the wave of social media. The third generation is all about mobile.

According to the prediction, one of the problems for Facebook is that the company has not been able to monetize its burgeoning mobile base. Company executives even raised the issue in their pre-IPO filings with the U.S. Securities and Exchange Commission this spring (2012). And that could really hold back the company and give others a chance to flourish in its place. The prediction is also based on the fact that Google has struggled with moving into social, and Facebook is going to have the same kind of challenges moving into mobile.

Facebook will not disappear in 5 years, even with the rise of more vertically oriented sites such as Pinterest (<http://www.pinterest.com/>) (Figure 4.6) and Goba (<http://goba.mobi/>)—consumers will still need a social home-base where all their acquaintances and friends are accessible.

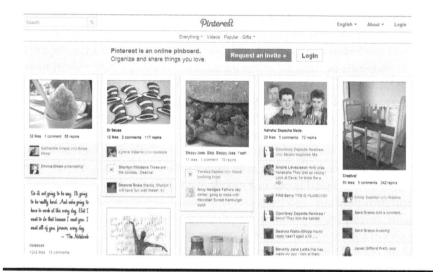

Figure 4.6   A challenger to Facebook.

If Facebook fades, it will be at the rise of other, more nimble companies; this is no different than the fading of AltaVista, AOL®, and Yahoo!. Consumer taste and technologies change, and if you are the 800-pound gorilla and miss a beat, you may be headed down. Facebook is making slow progress on the mobile front; the Facebook mob may find the interface difficult to see and use.

But Facebook has recently redesigned its mobile interface to make it easier to see pictures, for example, which is a step in the right direction. But they also need to build a mobile-enhanced business model in order to attract advertisers as well. CurrentAnalysis® (currentanalysis.com) agrees that mobile will be important—either as a platform that helps Facebook prosper or as a platform that causes it to stumble. According to the research firm, they believe it is simply a hugely important market driver, which will most likely play a significant role in the future of Facebook, depending upon whether or not the company is able to respond to this sweeping mobile trend.

However, with a Facebook mob of nearly a billion, data mining mobile devices are vitally important—as three in five mobile owners are accessing social media each month, according to comScore (comScore.com). Of course, with the increase in mobile social activities, Facebook and Twitter are experiencing a measurable influx in mobile activity in their marketplaces. Facebook's U.S. monthly mobile mob grew 50 percent in a year to 57 million, while Twitter's mobile user base skyrocketed 75 percent in 2012 to 13.4 million monthly users for the same period according to comScore (comScore.com).

But Facebook continued to beef up its mobile know-how by acquiring Pieceable Software, which makes an iOS app called the Pieceable Viewer; this app lets publishers view their iPhone and iPad apps through a Web browser. Facebook has also made several other mobile buys; it snapped up the mobile commerce start-up Karma, according to TechCrunch, as well as the developers of the Lightbox Android photo app.

The most predominant change in mobile social networking behaviors, according to comScore (comScore.com), is that more mobile users are accessing social media via mobile apps. More than 38 million people accessed a social networking site or blog via an app, representing a 126 percent change from a year ago (2011). But with 42 million U.S. mobile users accessing social media sites via a mobile browser, the browser still remains the most popular mode of social networking on mobile—for the time being.

Facebook offers the ability for direct advertisers and agencies to buy sponsored stories on mobile devices, exclusively. For those familiar with Google AdWords advertising, this new Facebook functionality is equivalent to mobile targeting for paid search campaigns. Sponsored stories allow Facebook advertisers to surface WOM recommendations about their brand that exists organically in the Facebook News Feed.

Sponsored stories are different from ads and can amplify the brand engagement of the target audience. For example, if a person's friends "Like a Page," in addition to seeing that news story in their News Feed, they can also see the same story on the right-hand column. Sponsored stories are available for ads that promote a Page, Place, app, or domain. The sponsored stories targeting options now available through the ads API and Power Editor include

- All placements: This option includes right-hand side, News Feed desktop, and News Feed mobile.
- All desktop placements: This option includes right-hand side and News Feed desktop.
- News Feed: This option includes News Feed desktop and News Feed mobile.
- News Feed mobile: This option includes News Feed mobile only.

What does this mean for Facebook and its advertisers? This change certainly provides marketers with more control over their ads and visibility into their data for optimizing campaigns. But to anyone familiar with advertising, this also means more ad revenue for Facebook. Consider these figures: Facebook made $3.2 billion in ad revenue last year, with $500 million of that revenue being mobile ads. With the floodgates now open for sponsored stories on mobiles and the continued growth of monthly active mobile users, it is clear that Facebook is set to generate hundreds of millions, if not billions, of dollars in revenue from this untapped mobile ad market.

Facebook may want to consider providing more visibility into the algorithm that influences the ads appearing in a user's news feed, such as click-through-rate. This algorithm might be similar to Facebook's existing EdgeRank, which factors affinity, weight, and decay into the ranking of social content appearing in news feeds. Developing a transparent and effective metric to promote ad relevancy would benefit marketers and Facebook users in a landscape of dynamically changing content and consumer intent.

Facebook provides the ability for advertisers to purchase Sponsored Stories in News Feeds specifically on mobile through its Ads API. Nanigans, an initial Facebook Ads API developer and Preferred Marketing Developer, developed and integrated this capability into its AdEngine™ platform. Nanigans customers can now select to deliver Sponsored Stories exclusively to Facebook mobile News Feeds. Over half of Facebook's 901 million active users visit the social network on mobile devices, and American subscribers alone spend, on average, more than 14 minutes a day on Facebook on their mobiles. The mobile marketing opportunity is impossible to ignore, and provides large-scale advertisers the ability for the very first time to reach Facebook users on-the-go.

This Nanigans mobile advertising feature is incorporated as a new targeting field in its Ad Engine. Marketers can indicate whether to deliver Facebook Sponsored

Figure 4.7   A "like" ad.

Stories to mobile News Feeds, desktop News Feeds, or both. Marketers can leverage Nanigans' advanced Sponsored Story targeting capabilities alongside mobile News Feed delivery. For example, marketers have the ability to target Sponsored Stories to Facebook users on mobiles based on their demographics, including gender, age range, and geographies.

Marketers can benefit from the Ad Engine automated optimization capabilities and robust reporting dashboard to understand how Facebook users are engaging with their brand or app in Facebook mobile News Feeds. The Ad Engine allows marketers to track and optimize their mobile advertising efforts on Facebook from impression and click all the way through the marketing funnel.

The Facebook marketplace is now the destination where people spend most of their time. A survey of U.S. Internet usage by Citigroup (citigroup.com) showed that Facebook claims 16 percent of people's online time, on average; that compares with 11 percent for Google and 9 percent for Yahoo! properties. It is clear that the level of engagement shows search Google and social Facebook are powerful channels individually, but that in combination, they create a virtual circle of knowledge and opportunity for the mobile miner and marketer. Facebook's mobile ads as reported by SocialCode (socialcode.com), suggests that sponsored-story "like" ads that appear in mobile news feeds get more clicks than the same units placed elsewhere (Figure 4.7).

The research examined more than 7 million Facebook impressions that were running fan-acquisition campaigns, including approximately 242,000 mobile news-feed impressions that generated 1,911 clicks. The click-through rate for mobile ads amounted to 0.79 percent, compared to the 0.148 percent average across all five placements studied: mobile, desktop news-feed only, desktop only (comprised mainly of right-rail ads), news-feed only (desktop and mobile), and the "control" group (uniform bids made across placements). The click-through rate for desktop-only news-feed ads falls roughly in the middle at 0.327 percent, according to SocialCode (socialcode.com) data.

Effective CPMs (cost per thousand ad impressions) for the mobile ads—which were bought on a cost-per-click basis, along with the other inventory in the research—were $7.51, compared to the $1.62 average across all five placements. SocialCode (socialcode.com) says the higher CPMs for mobile ads were largely due to the higher click-through rates they have been generating. If the marketer can make an ad that users "don't hate," it is actually extremely monetizable. A substantial number of clicks on mobile ads come from users who were fumbling with their mobile's keypads and clicked mistakenly, which would help explain why the "click-to-like" ratio, or percentage of users who clicked on an ad who subsequently liked the post or fan page, is lower than average for mobile.

The SocialCode (socialcode.com) results show that display and search ads on mobiles generally have higher click-through rates than desktop ads. Twitter also recently reported that engagement rates for its mobile-promoted tweets are higher than on desktop clients. However, there is also the argument that click-through rates are not really the right metric to be looking at anyway, as a recent study from the start-up Pretarget and ComScore (comScore.com) shows the correlation between that click and a conversion is virtually nonexistent. Even if mobile ads are nondisruptive and more native to the user experience, causing users to click on them more, the question remains as whether the Facebook mob can get the frequency right.

Facebook allows mobile developers using the company's API to cleanly integrate Facebook's iconic "like" button into their design, allowing users to perform the equivalent of a "like" within another app and then cross-posting that action to Facebook. Apps that require a user to rate something—such as a five-star rating for a restaurant, for example—will not be able to transform those actions into "likes." And apps that do take advantage of the "like" feature will require a user's permission to post that news to the person's Facebook news feed. This development represents just another step in Facebook's continuing integration into the mobile and social technology spheres. Just recently, Apple announced that Facebook would see integration with the iPhone in iOS 6, thus allowing users to more easily cross-post items to Facebook from their iPhones.

California Facebook users worried about privacy while using applications can breathe a sigh of relief. Based on an agreement struck in 2012 with the California Attorney General, all apps in Facebook's AppCenter are required to have a written privacy policy detailing what profile information is used and how it is shared.

4.7 The Google Mobile Mob

The Google Places platform offers up webpages dedicated to individual businesses with driving directions—showing the Google mob where they are located, along with street-level images on Google Maps and customer reviews of services and products—be it a pub, a restaurant, or a store. Businesses can also advertise through their Google Place pages; Google sells ads alongside search results, and firms that already show up on Google should still verify their listing and make sure its details are accurate and thorough.

Processing the mobile mapping request from users of all kinds of mobiles has provided Google with valuable insights into people's whereabouts and preferences. That, in turn, has helped Google sell more ads to local businesses, including more three-dimensional imagery. Google has 1 billion users and has provided 26 million miles of driving directions, and 75 percent of the U.S. population can see a high-resolution image of a home. Six years ago, that was 37 percent; the Street View feature has been enabled across 5 million miles, which allows areas to be seen at ground level.

Obscure areas will also get photos, such as the Grand Canyon, and the photos will be taken from equipment designed to be attached to a backpack. The equipment will be used on photo-capable bikes, instead of cars. Airplanes will also photograph cities to conjure more realistic 3-D views of metropolitan landscapes in the Google Earth™ version of its maps. The photos taken by airplanes are automatically converted into 3-D replicas.

The Google mob can start with Google Places; these Google listings are an easy way to maintain a mobile presence even if a business does not have a mobile site. Merchants and retailers can visit Google Places anytime to edit their information or see how many people have seen and clicked on their listing—an important metric for mobile analytics and marketing. Businesses can make their listing stand out with photos and videos; custom categories can be included to highlight the brands they sell and how to find parking; time-sensitive coupons or offers can be made to mobile users to encourage them to make location-based purchases.

SoLoMo (<http://www.slmtechnology.com/>) is an amalgam of social, local, and mobile, and this is the battleground where Apple and Google are locked in a fight, each trying to own more of the customer as the Web moves off the desktop, out of the office, and into peoples' pockets and purses. The first phase of the Internet was about constructing a virtual world that was isolated and apart from the real world. The next phase, driven by the move to mobiles such as the Apple iPhone and other devices, is about meshing together the virtual world with the real world.

Both Google and Apple already own the "Mo" part. The Google Android mobile operating system gets 900,000 activations a day. Apple has sold 365 million iOS-based devices, some 80 percent of which are running the latest Apple mobile operating system. But in order to mesh the two, both need two things: (1) geographical data to tell consumers where everything is, and (2) local data, the "Lo" to tell consumers about the neighborhood.

Google already has a mapping service; it acquired review site Zagat (<http://www.zagat.com/>) for $151 million to get into the local-review business, and rolled this out as Google+ Local. In addition, it will add reviews from friends on Google+ to consumer search results. Furthermore, Google has ramped up efforts to attract local businesses. Central to the effort is Google+, the company's social network, which it hopes consumers will use to interact with local businesses that now have special webpages on the network. Those Google+ pages will draw traffic from the company's Web-search engine. When shoppers visit these businesses, Google wants them to use their Internet-connected mobile like a digital wallet, earning loyalty points and making payments at stores that sign up for new Google new services.

Which leaves just the "Mo" and hence Google's relentless drive to get people to use Google+, a social service that recently had 170 million users, of whom 50 million visit every day. In contrast, Facebook, with which Apple appears to have mended its fences after an at-times fractious relationship, has more than 900 million users. After Apple's successful integration of Twitter into the last iteration of the operating system, it comes as no surprise that Facebook is integrated into the next phase.

However, by owning every part of the mobile, Apple users get a hugely consistent experience on their devices. The way the services all integrate with each other is unparalleled. An example of this is the new Apple Passbook app, which is likely to be big enough that any company using loyalty cards will need to consider integrating. For example, suppose a consumer is going to the movies. He can use Siri to find out what is on near him; then he can use iTunes, which has his credit-card details, to pay for it; and then he can use the new Passbook to store his tickets. He can use his iPhone to get him to the movie and perhaps NFC to redeem his ticket. The whole experience is seamless.

The Apple payment system is simple and quick, as anyone who has purchased goods on iTunes can attest. Developers like it, users like it. By contrast, Google checkout service has struggled. However, in raw numbers, Android is everywhere, but the reality is that the Apple demographics are highly attractive. So who are the big losers in all of this? First, the other mobile manufacturers: while Nokia has long identified mapping as a key part of its strategy, does it have the muscle to exploit it? The second losers are the network operators. For a long time, mobile network operators have dreamt of new revenue streams to replace the declining voice and text services. But it is difficult to see where they can jump in that chain. There are very limited opportunities for carriers to exploit the iOS ecosystem. And there are very limited opportunities apart from acting as a retail channel for the Apple mobiles.

Google Places asks businesses to verify their submissions by phone or postcard. Google does this to make sure that only the right people are able to change any public data about a local merchant or retailer. That business listing, also known as a Place Page, is a webpage to organize all the information for every imaginable place in the world. There are Google Place Pages for businesses, points of interest, transit stations, landmarks, and cities all over the world. Once a business has verified its Google Place listing, it can enhance the Place page by adding photos, videos, coupons, and even real-time updates such as weekly specials all on its Place page, to the moving Google mob.

Verifying the listing gives them an opportunity to share even more information about their business with the Google mob. Each business listing on Google represents a "cluster" of information that Google collects and organizes from a diverse group of sources, such as Yellow Pages, as well as other third-party providers and aggregators. However, the basic information that a business submits through Google Places is the information that Google trusts the most and is elevated and prioritized within the search engine. This means that it will appear instead of any basic information that Google aggregates from anywhere else.

To make sure the basic information a business submits is accurate, Google will ask the merchant or retailer to verify it first by entering a PIN that will be sent to either their business address or their phone number. This is an important factor for leveraging mobile search marketing for local merchants and retailers. Google highly recommends that these local businesses develop and use a mobile website as consumers are increasingly using mobiles in their research and shopping.

Writing apps for the iPhone involves a standard-size screen and, unless iPhone owners are slow to upgrade their OS, one version of iOS at a time. But writing and testing apps for Android and all its myriad devices can be a daunting task. It does not help that many Android devices are slow to upgrade to the latest version of the software: Only 7 percent of Android devices run "Ice Cream Sandwich," a.k.a. Android 4.0.

According to the app-analytics firm Flurry, developers build two iOS apps for every Android app they create. In return, they make four times as much revenue from iPhone users as from Android users. That discrepancy is crucial because, as Flurry (Flurry.com) noted, "If the developer community embraces one platform over the other, developers will build the software that infinitely extends the value of the consumer experience, giving a platform a meaningful edge."

And it is not just developers put off by the complexity of working with Android. While the iPhone has made inroads into the enterprise market—at the expense of the Research –in Motion Blackberry—a recent report by Gartner (Gartner.com) showed that Android is "severely limited" in large companies because of the work involved in managing these same variations in devices and OS versions.

Of course, Google makes most of its money from ads. Android is helpful insofar as it provides a ready-made platform for serving those ads. But Android seems to be peaking just as Google is facing new threats on other platforms. Apple is already using its own mapping app for iPhones, displacing Google—and integrating Facebook functionality into iOS 6, which could sideline Google efforts in social networking.

None of this means that Android is going away. IDC (IDC.com), which predicts a falling market share for Android, still sees it at the leading mobile platform for years to come. But there are enough clear challenges before Android to slow its impressive momentum of the past few years. If that happens, any Google plans to make money from mobiles users could hit a snag, and the company will have to get creative about the mobile Web.

Millions of people search Google Places via their mobiles every day for local businesses and services, and a user is more likely to click on a local map listing than any other Web listing because it is neighboring and trustworthy. Some statistics to keep in mind about Google, Google Maps, and Google Places are that 75 percent of all Web users look for services and products within an area close to their home or business: 65 percent of Google map searches contain a local reference, and 46 percent of all visitors who click on a listing found in Google Maps will purchase the product or service offered via their mobiles.

Adding a listing to Google Places is free, and Google does not accept payment to include particular listings or sites in their search results. However, Google does offer locally targeted advertising via their AdWords program. Another requirement of Google Places is that *every business listing must have a mailing address.* This is the physical address where mail can be sent to a business listing with Google Places. Another restriction is that *there should not be more than one listing per physical location;* even if a retailer or merchant has multiple locations, they cannot have two listings.

One of the advantages of mobile search marketing and Google Places is its incorporation of Google Maps, which is a free service that offers a wide range of functionalities. Google Places takes advantage of Google Maps by providing mobile users with several geo features for navigating toward its recommended business listing as part of Google searches. Google Maps also assists Google Places by providing public transit information, walking or bicycling directions, as well as estimated driving costs.

For example, the Google mob can find these routes using several different kinds of roads or paths, as provided by Google Maps. Biking directions are available for 150 U.S. cities. This is one of the critical advantages of mobile search marketing by Google; it has been developing these geo-data features for several years prior to the introduction of Google Places.

Google has effectively monopolized the mobile search market as shown in Figure 4.8. The Google mobile search market share was 96.9 percent in May 2012, according to Global StatCounter (<http://gs.statcounter.com/#mobile_search_engine-ww-monthly-200812-201205>). For comparison, Google's U.S. desktop search share was 66.7 percent, according to comScore (<http://ir.comscore.com/releasedetail.cfm?ReleaseID=682913>) and probably even higher overseas. Given that Google is the default search engine on Apple iOS and Android, which represent around 80 percent of the global mobile market; its dominance is not surprising, but it also provides some insight into the mobile ad market.

The majority of mobile ad revenues comes from search, which is really just an extension of the desktop. Many assume that mobile will be a huge new revenue stream for companies such as Google, but advertisers may just be shifting their resources to meet changing consumer mobile behaviors.

Google+ Local and Local Business Listing Optimization (LBLO) training are being offered by Maximize Social Media LLC (<http://maximizesocialmedia.com/>). This social media management company will provide insight into the latest change in the Google business listing directory. Maximize executives will offer insight into how businesses can optimize their listing placement and integrate

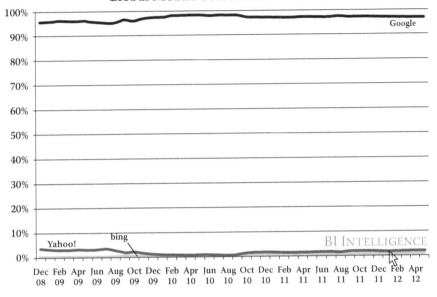

Silicon Alley Insider  Chart of the Day

Global Mobile Search Market Share

Figure 4.8   Google totally dominates mobile searches.

social successfully. LBLO and social are converging, just as SEO (search engine optimization) and social converged with Google and Bing integrating social preferences within the search engine results page. This is great for businesses that have been engaged with their customers using social media, but a lot of businesses are now playing catch-up.

Increased access by mobile users derailed the ability of Facebook to earn revenue from those users. It brought to the forefront of popular press the nagging question lurking in the minds of companies attempting to grow their mobile businesses: How will mobile successfully make money? Can it be monetized at all? *Selling apps is a first step.* Apple and Google host robust app stores and others are following suit, such as Amazon and Facebook. The majority of apps offered by Apple are for sale, while most Google apps are free.

Today, the average price for an app for a mobile is $3.77, and the app stores take a 30 percent cut. Apps for tablets generally have a higher price tag. The other obvious way to make money with apps is to offer "brand" apps to sell their wares as another venue for the existing brand, such as Starbucks and Amazon. Both have had tremendous success in offering apps as a mobile purchasing agent. For example, Starbucks launched its app, which has been downloaded 26 million times and accounted for $110 million in Starbucks card reloads.

However, mobile offers much more potential and an opportunity to create a new business model. Mobile has more going for it than the desktop did a decade ago.

1. Users have been through tech cycles now, so the experience is not entirely new—just a different venue.
2. Each new cycle is more quickly adopted, accepted, and embraced than previous cycles.
3. Advertisers and marketers are much more sophisticated and flexible, more willing to be innovative and creative toward working with a new venue.
4. Mobile grabs and holds a user's attention instantly.
5. Mobile requires more focus to hold, to change content, and to engage.

A mobile user is generally not at home or in an office and thus there is a greater need to ignore the surrounding noise and focus on the device. Often, the user's goal is to focus on the mobile, and not the environment, such as when in line, shopping, or on public transportation. Moreover, switching between content on a desktop is easy and requires just typing in a new URL, or clicking on another open browser. In contrast, on a mobile device, every move requires "backing out" of the previous app before selecting a new venue.

Also, consider the data. Flurry Analytics (Flurry.com) found in a recent study that for the first time, minutes spent on mobile apps surpassed that on the Web. The data points to a continuing trend as one in which mobiles will exceed those of desktops and notebooks. One can argue that not only is the quantity of time spent on mobile bypassing the desktop, but so is the quality of time.

Ooyala tested video engagement on various devices and found that mobile viewers have more patience to watch videos on their mobile device than on the desktop (<http://videomind.ooyala.com/blog/size-matters-consumers-watch-40-more-video-tablets>). In fact, twice as many mobile viewers (40 percent) were likely to watch three-quarters of a video compared to desktop viewers. For medium and longer videos, viewers are watching from mobiles. Most videos on desktops are under 3 minutes, while the greatest percentage of videos watched on a mobile are longer than 10 minutes.

Apps deliver information to users. Mobile users can download apps so information that they are interested in is delivered to them, rather than having to type with big fingers on tiny buttons in a super-small search box. So, if someone is interested in décor, they can download Houzz (<http://www.houzz.com/>); or if they are interested in ski conditions, they can download SkiReport (onthesnow.com); or MyFitnessPal (<http://www.myfitnesspal.com/>), to track calorie intake.

Users can find the most suitable apps for their topics of interest, and that data is delivered directly to them, as conveniently as a touch of an icon. Apps are a tremendous venue to deliver targeted information because the interest is defined and the user has "opted in" to receive information—versus search, which could potentially be one-off or random. Apps provide the opportunity to engage with a potential customer, rather than simply flash a message.

The obvious limitation with traditional Internet advertising (banner ads) on a mobile device is the small screen size. The obvious advantage of mobile is that it is more engaging and targeted, and through certain types of apps, the audience is well defined and open to marketing information. The challenge for marketers is to customize the format of advertising away from banners and develop toward well-placed demonstrations, introductions or tags, and the opportunity for a direct one-click sale.

Google introduced a service for businesses to help companies organize mobile workers, share location data, and collect work-related data on-the-go. The app, called Google Maps Coordinate (<http://www.google.com/enterprise/mapsearth/products/coordinate.html>), provides real-time information of workers on location. The mobile and Web-based versions of the app will allow organizations to update and track the progress of various tasks assigned to workers on-field. Google stated via a company blog post that companies can use the service for $15 per month to share location information and also maintain a permanent record of where everyone has been. It combines the power of Google mapping technologies with mobiles to help organizations assign jobs and deploy staff more efficiently.

Once employees download the app on their mobile after agreeing to the terms and conditions, they can update their location every 5 seconds or at 1-hour intervals. Employees can also schedule their tracking with the Android app. They can choose to be invisible on the map or set it in a manner that allows them not to be tracked after work-hours. Google said beta testers include a utility company, a city government, a pizza shop, and a large-scale telecom provider.

People use their mobiles to discover and engage with the world around them. Searching for local information is one of the most common activities on mobile; in the United States, 94 percent of respondents said they had done so, and nearly every surveyed country reported numbers over 80 percent (Google.com). With Google Maps, they make informed decisions about places to see, shops to visit, meals to eat, and more. To enable advertisers to better connect with potential customers via mobile search, they have redesigned their local ad formats for Google Maps for Mobile (<http://www.google.com/mobile/maps/>). These have produced measurable results: in initial tests, these redesigned formats increased click-through rates by 100 percent. This visual redesign will be rolling out to Android phones.

When Google introduced Google Maps, most people quickly realized that the days of the stand-alone GPS device were numbered. The Google Android OS offered maps, turn-by-turn directions, even traffic—all for free. But Google has inexplicably ignored a central feature of GPS devices: the home location. When setting up a stand-alone GPS from Garmin (<http://www8.garmin.com/apps/>) or TomTom (<http://www.tomtom.com/en_us/>), the first question users are asked is to set their home location. Not so with Google Maps; in fact, Google took 3 years to implement a home feature on the desktop version of Google Maps.

Android is a vital component in the mobile Google marketplace, and its plan is to bring in the big dollars; and given the projected growth of both mobiles and usage over the coming years, it is going to get a lot more important. Google now has activated some 190 million Android devices worldwide. That is a big number, but a far bigger one is the $2,500,000,000 that mobile users are generating for Google yearly—a figure that has gone up 2.5 times over the past 12 months—and is anticipated to double to $5 billion over the next 12 months.

Google does not charge a licensing fee for the Android OS, and does not keep a cut of the Android Market sales; 70 percent goes to the developers, with the remaining 30 percent split between the carriers and payment processors. However, it does offer a great platform for shepherding users to Google properties where they can look at their ads. And then there are apps that can also be home to mobile ads thanks to Google purchasing AdMob, a mobile ad network, for $750 million.

In a case study, RingRevenue® (ringrevenue.com) (<http://blog.ringrevenue. com/bid/64084/Search-Marketers-Want-to-Improve-Mobile-CTRs-by-250>), a call performance marketing firm, revealed results from a recent test: mobile search ads with phone numbers outperformed mobile search ads without phone numbers by nearly 250 percent. Mobile search ads with phone numbers simply perform much better and drive more valuable conversions for mobile marketers. According to Google, 70 percent of all mobile searches result in an action within an hour. Combine that with adding campaign-specific phone numbers to mobile search ads and it should be a no-brainer for mobile analysts and marketers. Clicking-to-call is one of the clearest and most intuitive calls-to-action for mobile consumers—where calls to local businesses have been shown to result in higher conversion rates and average order values than stationary clicks.

4.8 The Twitter Mobile Mob

It is an epic battle between search marketing (Google) and social marketing (Facebook) for the estimated $40 billion of small and medium businesses in local advertising for the current year, which is increasingly mobile. Not wanting to be left behind, the micro-blogging giant Twitter has also launched its own Twitter Places, which allows devices to broadcast their "tweet" and their location—including the businesses they are visiting—to all the followers of their messages.

Twitter has revealed that the site generated more advertising revenue from its mobile platform than from its website. The Twitter mob has a long heritage in mobile—the 140-character limit was driven by the 160-character limit of text messages—and its management clearly wants people to understand that it does not see the same challenges in monetizing mobile as Facebook does. This is not to say that Twitter does not have challenges of its own in mobile. In contrast to Facebook, which has not found a way to monetize the use of its own clients on mobiles, the biggest challenge for Twitter will be finding a way to monetize the use of third-party clients on mobiles, which were the only option in the early days and are still very popular. Twitter has tried to overcome that problem by acquiring several of the most popular clients, but it still does not have a definitive strategy for monetizing mobile either.

Twitter currently has 140 million active users who produce 400 million tweets a day. Two and a half years ago, the company decided to go with an ad-based model, although it also pursues data licensing opportunities. The company clearly has a solid business model, while also providing a wonderful service that brings the world closer together. The company appears to have a very bright future, and it may be one of the leading pioneers in figuring out how to be profitable in the mobile era.

Twitter has rolled out a program that lets users create trends based on whatever location they choose, as well as based on who they follow on the short-messaging blog. Their Twitter Trends help users discover the emerging topics that people are talking about on Twitter. The user can see these topics as a worldwide list, or can select one of more than 150 locations. In order to show emerging topics that matter more to the user, they have been improving their algorithms to tailor Trends based on a user's location and who they follow on Twitter. The trends lists will work on Twitter's website, as well as with Twitter mobile apps for the iPhone and Android mobiles. If users do not want these "tailored trends," but prefer a general list of trends, all they have to do is change their location on the Twitter site. Of course, anytime they want a glimpse into the conversations that are happening on Twitter around the world, they can always select individual countries and cities to see Trends in those locations.

The Twitter "Tweet with Your Location" feature allows users to selectively add location information to their Tweets. *This feature is off by default, so users need to opt-in to use it.* Once they have opted-in, they will be able to add their location information to their new Tweets on Twitter.com and via other mobiles and apps that support this feature. Some third-party apps will let mobiles tweet broadcast their exact address or coordinates. Tweeting with a place can add context to updates and help users join the local conversation, wherever they are.

Now, when a mobile Tweets with its location, they can specify an explicit place or other point of interest. In this way, mobiles can provide additional information that makes their Tweets more meaningful—without taking up extra characters. The publicly shared location information will be either the mobile's *exact location* via its self-provided coordinates—store, mall, bar, etc.—or a neighborhood or a section of a city.

Tweeting via Places allows a mobile to add information to its updates and engage in local conversation about businesses wherever they are. Twitter Places has integrated other location-based social networks, such as Foursquare and Gowalla. Many Foursquare and Gowalla users publish check-ins to Twitter Places, with physical location being the key component of these Tweets.

This means that when a mobile clicks on a Twitter Place, it will see standard Tweets and check-ins from Foursquare and Gowalla. The use of a Twitter advance search by businesses can immediately limit a mobile's search for specific Tweets within a geographic area and refine it further by searching for other specific factors such as keywords, people, places, dates, reviews, and attitudes.

Twitter recently released an update to its expanded Tweets, which enables people to preview articles, video, and other content directly on Twitter.com or on the mobile Twitter site. The change means that more content will be available directly on Twitter without users having to leave the website. Twitter partners include *The Wall Street Journal, Time,* MSNBC, *The New York Times, The San Francisco Chronicle,* and *DerSpiegel Online.*

Previously, Tweets from services such as Instagram (<http://instagram.com/>) or YouTube showed photos or videos from those services directly on Twitter, but now news publishers will also have this functionality. When users click to expand a news Tweet, they will see the first portion of a news article as well as the writer's Twitter handle. For images, users can see content within the Twitter stream from BuzzFeed (<http://www.buzzfeed.com/>), TMZ (<http://www.tmz.com/>), or WWE (<http://www.wwe.com/>), and video will be available from BET 106&Park (<http://www.bet.com/shows/106-and-park.html>), Lifetime® (<http://www.mylifetime.com/>), and Dailymotion (<http://www.dailymotion.com/us>). This is just about content at this point, but it is not difficult to imagine this in-stream content being made available to advertisers. Currently, this is only available in the browser, but it will soon come to iPhone and Android native apps for the Twitter mob.

Twitter is offering more rich media and content previews to Tweets viewed on its mobile.twitter.com website and on mobile. The new content previews will be displayed on expanded Tweets with links to articles from *The New York Times, The Wall Street Journal, Time,* and others. Users will be able to view the article's headline, introduction, and sometimes the Twitter accounts of the author and publisher. Tweets with links to other sources such as BuzzFeed and TMZ will, in addition, display photos and graphics.

Twitter lets their mob create trends based on whatever location they choose, as well as based on who they follow on the short-messaging blog. Trends help their mob discover the emerging topics people are talking about on Twitter. Users can see these topics as a worldwide list, or select one of more than 150 locations. In order to show emerging topics that matter more to them, the company is improving their algorithms to tailor Trends based on their location and those who follow them on Twitter.

Twitter was born from mobile and grew up in mobile, and it is influencing the way mobile is developing. On the other hand, Facebook has always been a desktop experience. Its mobile app is pretty terrible. It is an over-stuffed piece of software that attempts to cram in every one of Facebook's myriad features. The experience is slow and cumbersome.

Twitter, however, is light and speedy and simple. People are constantly closing Facebook on their mobiles and switching to Twitter, simply because they are tired of waiting for it to load. It is perhaps for this simple reason that more people are starting to post their status, location, photos, and questions to Twitter—it is clean and simple and works so well—because that is the promise of the 144 characters Twitter presents on its surface.

Research from Localytics (Localytics.com), a mobile analytics firm reveals that mobile users spend the same amount of time with news apps as they do with the Twitter mobile app—roughly 115 minutes each month. It is inevitable that in just a few short years, mobiles will be the most common form of accessing the Internet. So where do traditional local news stations stand throughout the mobile transformation? Localytics data showed that mobile users spend two-thirds as much times using apps for entertainment, health and fitness, and sports; while less than half as much time is spent using games apps or news apps.

There is clearly a lot of potential for local news apps and plenty of room for growth in the industry as a whole. Both tablet and smartphone news apps could be advantageous as the vehicle for delivering information to people. Although promising for consumers, would these mobile apps really drive in the dollars for the news organizations? Not necessarily, considering that the apps would not be a direct source of print or broadcast.

News aggregator apps, Flipboard (<http://flipboard.com/>) and Onswipe (<http://www.onswipe.com/>), seek to "repackage" already-available content to then make it simpler to access through mobile browsers. However, the companies that manage the tools are the ones that decide which ads to promote, while publishers who distribute their content through the news tools only earn a slight portion of the generated ad revenue. Most news brands that offer their own mobile apps through mobiles serve ads from revenue-sharing mobile ad networks.

CNN recently purchased the popular news aggregator tablet app Zite, which seeks to eventually support the news organization directly via Apple mobiles. Although it would greatly benefit the large CNN corporation, it would have little effect on local news station outlets throughout participating cities. Although the fate of community news stations cannot fully be determined, their predicted pathway is quite unsettling.

Local news outlets can sit still for now, as the desktop is currently the primary source where people retrieve most of their digital news, as the majority of people say that they prefer to get their news via desktop direction to a news organization's website. But with mobile apps on the rise, there is no doubt that mobiles will achieve a sizable place in news consumption in the years to come—and become a new target for data mining mobile devices.

While Facebook is getting a lot of attention since their IPO, Twitter has been quietly making changes for the future with hardly any notice. The company hopes to become a social media search engine the way Google is for the Internet. To achieve that, Twitter has announced that they are working to improve search. With so many companies now using Twitter for corporate communications, this evolution is designed to drive their advertising and market value.

In tangent with search, the company also rolled out its "expanded" Tweets. The Twitter mob can now preview a page linked to a Tweet without having to load it. This makes Twitter faster to use, and is particularly useful for mobiles. The company wants to position itself as the "go-to" site as mobiles overtake the desktop. Currently, about 48 percent of social media use is mobile (Localytics.com), of which Twitter is keenly aware.

Another change is something that Twitter calls "tailored trends," which chooses specific trending topics depending on the user's location and who they follow. The feature offers the Twitter mob the choice of 150 cities globally as their home location. Twitter will likely become highly segmented by people's interests, topics, and followers in the future. A Twitter feed is a commitment that requires planning for content and the data mining of mobiles.

The Twitter mob can burn through a lot of content at 144 characters a post, and they do not want their Twitter feed to become stale. The user might want to be on it just to receive Tweets from other users. The Twitter mob provides a great a news feed and an effective way to keep up on their industry. A new study from the Pew Research Center (Pew.com) provides some insight. According to their report, about 24 percent of all Internet users are now on Twitter. Not surprisingly, the most active demographics by far are urban residents ages 18 to 29 years. Minority populations also tend be higher on Twitter, which is not the case on Facebook and LinkedIn.

Whether the user is on Twitter or thinking about it, keep in mind that creating effective, engaging content is an art form. A good way to get started is to follow organizations that are doing it well—and have the followers to prove it. They need to learn how organizations use questions, varied sentence styles, hashtags, and links to create engaging content. Also, the developer and marketer need to pay attention to what Tweets attract attention as they scroll through the feed. The Twitter mob started out as a small competitor to Facebook, but like the Tortoise and the Hare, it may just end up becoming very important in a mobile-dominated world.

Local businesses can leverage Twitter Places with its unique real-time location feature. For example, a merchant or retailer can search by keywords, places, dates, or attitudes to find Tweets they may want to respond to via Twitter. A local tweet searching for "tacos" or "enchilada" might prompt a Mexican restaurant to post its lunch special, complete with photos. And as their customer fan base grows, the restaurant can ask Twitters for suggestions on what they would like to see on their daily special. The core of social mobile marketing via Twitter Places is customer engagement and participation.

An advanced search will bring up all the recent tweets that meet a business's criteria with the handle and avatar of the person who tweeted. From the results page, a business can follow these people or click on their profiles for more information, with the probable results of harvesting new followers and customers. Twitter has released an API (application programming interface) that lets developers integrate Twitter Places into their mobile apps. Twitter, of course, supports all the major mobile browsers, including Apple and Android.

Twitter Places not only knows who the users are, but also what they are interested in and where their mobiles are and when. Using text analytical algorithms, Twitter can readily cluster the similarities of desires and preferences of mobiles who favor certain types of services and products, which they can monetize via mobile analytics and mobile marketing. With Twitter Places, the firm is attempting to organize and provide advertising based on the physical location of tens of millions of local businesses via mobiles. Here are some tips for marketing to the Twitter mob:

1. Do some tweet watching. This is something really simple. The mobile marketer needs to go to the store, go to the mall, go out to dinner, and sit back and watch people tweet. Just watch. This is old-school marketing, and they should see that everyone is on their mobile, they are spending quite a bit of time on their devices, they are not necessarily making phone calls, and they are not just doing short message service (SMS)—they are doing a lot of things.

   When the developer and marketer see that happening around them in the Twitter mob and everywhere they go, how they interact, they need to think about how they can insert their product, brand, or message into these experiences. They need to look at the world around them and listen to young people who grew up in the digital age. Their digital behaviors are completely different. The marketer will see that this is definitely the route to take: invest the time to learn it and understand it—marketers need to explore this new Twitter mob.

2. Mobile marketers need to keep an eye on what "the little guy" is doing, as well as what the titans are doing. There will be a revolution of incremental innovations that are about to take place. It is so easy nowadays to build a mobile site, or an app, so what is happening is that a bunch of people are starting to solve the problems that they have been having in their own lives and industries.

3. Build a marketing program around tactics that make sense for the brand. Many are overwhelmed by Twitter, Facebook, Foursquare, and the like. But when talking to clients about creating content that provides value to existing and potential customers, they get that. Focus on the fundamentals. Do not pretend to be something you are not—and do the right thing!

4. Create interactive "rich media" designed specifically for a mobile site. On mobile, you do not have the flexibility but you have the entire device. You can incorporate rich functionality where you can shake the mobile or where it is actually using the camera function in augmented reality or the location function. There are so many other vectors or parameters that are unique to mobile that make it extremely unique, new, and rich.

5. Find out how customers want to communicate with a brand or enterprise. So texting is not replacing the Web. The Web did not replace the phone call. People who want to call will still want to call. People who never called before hopefully will engage through text. Some people prefer Twitter or are on Facebook all day. This is the age of choice.

4.9 The Amazon Mobile Mob

Amazon CEO Jeffrey Bezos thinks of their marketplace as a *service* that merges hardware, software, and the largest retail store on the planet. Their Kindle Fire mobile is, of course, the hardware, but really, it is their site—and more importantly, their content. It is the seamless integration of a mobile, cloud computing, and the largest retailer in the world. The Amazon marketplace contains multiple warehouses packed with more than 18 million consumer goods, e-books, songs, movies, television shows, and cloud storage space. The access to content is important—as Amazon transforms its business into a digital retailer and responds to consumer demands and their desire to shop via their mobiles.

At the core of the Amazon marketplace is its Kindle Fire (Figure 4.9), a mobile designed for consuming—which will provide direct access to all types of products, services, and content—with the ability to enjoy digital video and music. Most importantly, Amazon already knows what consumers like to read using its proprietary collaborative filtering technology and can now expand this to include other types of retailing and digital products. The Kindle Fire runs a special Amazon version of Google Android OS; the user interface, however, eschews Android in favor

Figure 4.9 The Kindle Fire mobile.

of an Amazon-centered experience, built instead around Amazon Cloud Player™ and Amazon Cloud Drive™.

The browser is called Amazon Silk™; it uses the Amazon EC2™ computer cluster. This means that Amazon will employ huge processing power (compression) to reduce bandwidth demand on the mobile and put more processing in the cloud; incredibly, Amazon Silk learns a user's browsing patterns and preloads the pages they read the most. There is also a tablet-optimized shopping app on the device; this simplifies and streamlines pages, making it easier to buy products. The Amazon Kindle Fire is the key for an intuitive and seamless browsing and shopping experience for the Amazon mobile mob.

Two-thirds of U.S. adults now use at least one mobile media device in their daily lives, according to a national survey recently conducted by the Reynolds Journalism Institute (RJI) at the University of Missouri School of Journalism (Missouri.edu). The RJI survey found that news consumption ranks fourth among reasons people use mobile devices, behind interpersonal communications, entertainment, and Internet usage for information not provided by news organizations. Despite the large number of mobile users, the RJI survey found that mobile news products do not appear to be replacing printed newspapers as quickly as was earlier predicted.

The increased use of mobiles does not yet appear to have accelerated the switch from print to digital news consumption as previous surveys have suggested. Forty percent of mobile users indicated in the RJI survey that they still subscribe to printed newspapers and news magazines. This percentage was almost identical for non-users of mobile devices.

The RJI survey interviewed more than 1,000 randomly selected respondents. The survey divided mobile media devices into four categories: large tablets, small tablets, e-readers, and mobiles. More than 21 percent of the respondents said they now use large tablets, a category of mobile devices that entered the market just 2 years ago. Results showed that Apple is dominating the large tablet market, with more than 88 percent of large tablet users owning an iPad, while Amazon is dominating the small tablet and e-reader markets. The survey also showed that mobiles and large media tablets are the two most popular devices for consuming news. News organizations should consider these numbers when targeting their audiences.

The Amazon and Apple mobs have built quite a bit of brand loyalty from their customers. Forty-four percent of Apple iPhone owners also own large media tablets, 96 percent of which are iPads. This obviously poses a significant challenge for publishers and advertisers who are trying to circumvent Apple by focusing their attention on mobiles with the Android operating system. These results suggest that to reach the highest percentage of mobile and large media tablet owners, news organizations must make their content available on Apple iPhones and iPads.

Concerning Amazon, the RJI survey showed 22 percent of Kindle e-reader owners also own a small media tablet, of which 71 percent were Kindle Fire tablets, while only 14 percent owned Barnes & Noble Nook tablets. The survey also found that Apple iPhone and iPad owners tended to be somewhat older and have significantly higher household incomes than owners of mobiles and large media tablets powered by the Google Android operating system.

Kindle Fire will have access to 100,000 movies and TV shows, 17 million songs, Kindle books, and "hundreds" of magazines and newspapers. Amazon Prime™ members enjoy instant, unlimited, commercial-free streaming of over 11,000 movies and TV shows at no additional cost. There will also be free Amazon Cloud Storage, so that the user can start a movie on Kindle Fire and transfer it to other devices, such as their TV or desktop.

Amazon recently released its long-awaited Cloud Player app for the iPhone and iPod touch. The free app allows users of iOS devices, including the iPad, to stream or download music stored online in their Amazon Cloud account. After launching its cloud drive and music service, Amazon has certainly taken its time before releasing a native app for iOS. Amazon made its Cloud Player compatible with the mobile version of Safari. Although a native app, Cloud Player also lets the user manage and create playlists, and play music already stored on their mobile. Amazon customers get 5 GB of free cloud storage and can buy additional space as well, including 20 GB for $20 per year or 50 GB for $50 per year. Users who buy a storage plan receive unlimited space for MP3 and AAC (.m4a) music files at no additional cost.

The Amazon Appstore Developer Program enables mobile app developers to sell their apps on Amazon.com. The Amazon Appstore currently supports the Android operating system and works on Android devices running Android OS 1.6 and higher. Amazon pays developers 70 percent of the sale price of the app or 20 percent of the list price, whichever is greater. Amazon is offering their Amazon Digital Services (AWS™) and their Web services to app developers.

For people who cannot stop hitting the "Add to Cart" button anyhow, Amazon has developed its own Amazon Mobile app for Android devices, making shopping on-the-go and maxing out on that credit card even easier. The latest version of Amazon Mobile brings along with it the following new features:

- Brand-new ways to sort and filter search results
- Faster, easier navigation with a cart shortcut and drop-down menu of the most popular pages
- Launch barcode scanning directly from the home screen
- Add items to any existing wish lists
- Additional country support: shop Amazon across eight countries, including Spain and Italy
- Bug fixes and performance improvements

The Amazon Mobile app on Android is still intact. Searching and comparing prices are a breeze, as is reading reviews. Wondering if that product a user is eyeing in the supermarket is too expensive, the user can use the app to discreetly scan the product's barcode and instantly know how much it sells for on Amazon. The

Amazon Mobile app also gives the user access to their existing shopping cart on Amazon, wish lists, payment, order history, 1-Click™ setting, and more. All in all, the app is as close as it gets to the full Amazon shopping experience.

The Amazon Appstore in the United States has grown to tens of thousands of apps and games in just its first year. As a result of features such 1-Click purchasing and Test Drive—which allows customers to try apps before purchasing them—developers report strong monetization from the apps they offer through Amazon. Amazon recently introduced an In-App Purchasing service that is simple for developers to integrate and helps monetize their apps and games even better than before, while still offering customers a seamless and secure 1-Click purchasing experience for the Amazon mob. Many developers have seen revenues skyrocket as a result of Amazon In-App Purchasing and are excited to offer their apps to even more customers outside the United States.

Amazon gamers are hungry for great content and are willing to open up their digital wallets to pay for it. Taking a successful platform such as that of Amazon and expanding it across the globe is going to give them an even broader customer base and create an opportunity to generate even more revenue. Developers can visit the distribution portal to learn about localizing apps with localized resource files for different regions, as well as other tips on how to prepare their apps for international distribution.

Developers have the ability to select the countries where they would like to sell their apps. They can also set their list prices for different marketplaces. Those already participating in the program will automatically have their apps made available for sale internationally by default. And developers can easily change international availability for their apps via the distribution portal if they do not wish to sell in select countries. Developers who are new to the program can sign up at the Amazon Mobile App Distribution Portal (<https://developer.amazon.com/welcome.html>).

Amazon Web Services (AWS) delivers a set of simple building-block services that together form a reliable, scalable, and inexpensive computing platform "in the cloud."

Mobile app developers can access these "compute-and-store" resources through simple API calls and with the use of the AWS Mobile SDK for Android (and other OSs). The SDK is a software development tool that makes it easier for developers to access AWS resources directly from their development environment. The AWS SDK for Android provides a library, code samples, and documentation for developers to build connected mobile apps using AWS.

The decision by Amazon to price the Kindle Fire much cheaper than the iPad 2 and substantially more affordable than the vast majority of Android tablets is a tactical move. The Amazon strategy is clearly not to make money on the sale of the mobile, but instead on the sales the device will generate in content as well as its thousands of products and cloud services. One of the key components in the Kindle Fire sales pitch is the integration of Amazon's many services, including its

Cloud Drive, MP3 download store, and Kindle e-books. Only Apple and its iTunes platform can equal Amazon integrated offerings. However, the Amazon marketplace sells more than digital products; they offer everything from shoes to diapers to cloud space.

Other mobile makers have gone for larger tablet space, thinking it would be more appealing to users; but the Kindle Fire with its seven-inch multi-touch display is a pocket-size device *designed for shopping*, and may prove that a bigger tablet space does not necessarily mean a better device. Amazon has designed the Kindle Fire to allow for the millions of people who visit its marketplace to buy more goods in a highly targeted, modeled manner: their in-house mobile analytics will make product recommendations that will be highly relevant to consumers and highly profitable for Amazon.

Google began selling its first tablet for $199, hoping to replicate its mobile success in a hotly contested market now dominated by the Amazon Kindle Fire and Apple iPad. By taking a greater role in the tablet market, Google hopes to ensure that its various online services remain front-and-center to consumers amid a changing technology landscape in which tablets by Apple and Amazon are increasingly becoming gateways to the Web and Web-based content such as movies and music.

Google's maiden entry into the tablet market will also see the advent of Microsoft Surface™. These mobiles could also help accelerate development of tablet-specific applications for its Android operating software, a key factor that has helped popularize the Apple iPad. The "Nexus 7" tablet was built by and co-branded with Asus of Taiwan (<http://www.asus.com/Tablet/Nexus/Nexus_7/>).

Google acquired its own hardware-making capabilities with the $12.5 billion acquisition of mobile maker Motorola Mobility. But Motorola, which Google has said it will run as a separate business and Google Glass™, a futuristic-looking eye-glass-computer that is capable of live-stream events, recording, and performing computing tasks will be used to develop hardware. The mobile will be available to U.S.-based developers for $1,500. The Nexus Q is a $300 device with a built-in amplifier that lets users stream content from Android devices onto their TVs. The Nexus 7 tablet, with its $199 price tag and seven-inch stature, is aimed squarely at the Kindle Fire, but the Nexus has a front-facing camera while the Amazon tablet does not.

Analysts consider the Kindle Fire a window into the Amazon.com trove of online content rather than an iPad rival, given the $499 that Apple asks for a device with a "retina" display that far outstrips it in terms of resolution. Google can similarly use the Nexus 7 to connect to its own online offerings, which include YouTube and Google Play (<https://play.google.com/store>), the name of its online store where it sells digital music, movies, and games. It will go after more cost-conscious users who might shun the pricier iPad.

Nexus 7 is an ideal device for reading books; Google will offer buyers of the Nexus 7 a $25 credit to spend at the Google Play store; and it showed off several media-centric capabilities, such as a new magazine reading app. They all but called it a Kindle Fire killer. They are clearly gunning for that No. 2 spot behind the

Apple iPad that is currently occupied by Kindle Fire. But the con is they do not yet have a footprint in people's minds and wallets as the go-to place to purchase and consume media.

Google has partnered with mobile makers to develop Nexus-branded mobiles for several years, providing a showcase product that delivers the Google ideal vision for a device based on its Android software. Extending the Nexus concept to tablets should similarly establish a model that other hardware makers can emulate, resulting in a more a competitive and uniform line of Android tablets to market, say analysts. The Nexus will feature the new 4.1 "Jelly Bean" version of Google software, as well as a front-facing camera, a 1280×800 resolution screen, and an Nvidia® Tegra® 3 processor. Google's free Android software is the No. 1 operating system for mobiles, with about 1 million Android devices getting activated every day.

However, Google has struggled to compete with the Apple iPad in the marketplace for tablets, largely because it lags far behind Apple and Amazon in terms of available content and tablet-specific applications, such as games. Meanwhile, Apple has increasingly moved to reduce its dependency on Google services on its devices. The Amazon Kindle Fire, while based on Google's open-source Android software, features a customized interface that does not use many Google services. The new software delivers faster performance, according to the company, and new features such as "voice search."

That range of services will be the secret to stitching together this rag-tag fleet of Android gadgets into a platform that can compete with Apple for minutes of users' attention rather than premium device dollars. The tablet's limited availability— executives said they had no plans yet to expand distribution beyond Google's own site—may curtail initial sales growth.

Google briefly sold a specially designed Android mobile, the Nexus One directly to consumers, but closed the online store after four months, saying it had not lived up to expectations. But it is the lack of "native" applications—that is, software designed with a larger tablet in mind—rather than ported from mobile that is the biggest impediment for Nexus right now.

4.10 The Mobile Fab Four

All major national carriers—AT&T, Sprint, and T-Mobile—sell their mobile data; they know what sites their mobiles visit, what apps they download, what content they like, and most importantly, when and where that mobile is. Verizon is offering customers' device location data and browsing history to third parties, combined with other anonymous demographics such as age and gender, thus enabling miners, marketers, retailers, and enterprises to conduct highly precise and relevant mobile analytics for modeling consumer behaviors. This kind of mobile data is very useful and lucrative for the triangulation of a device's time, location, interests, and anonymous demographics.

Figure 4.10   AT&T mobile data.

The AT&T AdWorks program promotes this data aggregation for mobile marketers with customized audience segments based on anonymous and aggregate demographics (Figure 4.10). Sprint, like Verizon, tracks the kinds of sites users visit on their mobiles as well as what apps they use. Sprint also uses their data to help third parties target relevant ads to its customers.

Verizon lets advertisers target customized messages to Verizon subscribers' mobiles, but it does *not* incorporate its customers' Web-browsing or location data. Verizon relies on other personal information, including customers' demographic details and home addresses. T-Mobile collects information about the sites that customers visit and their location; it uses that information in an anonymous, aggregate form to improve their mobile services. Apple, Google, Research in Motion, Microsoft, and Nokia also collect data about a user's device location in their servers via a combination of GPS and/or Wi-Fi network triangulation.

The rise of mobiles has given mobile providers and manufacturers an accidental treasure trove of marketable data. The mobiles in billions of purse or pocket gadgets are hyper-personalized tracking devices that know more about their human owners than they themselves do—what they want, when, and where. It is a situation in which the consumer *is* the device. There is a lot of money to be made of these largely untapped local and brand-moving marketplaces via the data mining of mobile devices.

A BIA/Kelsey (BIAKelsey.com) study predicts that U.S. local mobile ad revenues will reach $42.5 billion annually in 2015. Google, Apple, Amazon, Twitter, and Facebook are all scrambling to sign local businesses to their new marketplaces. Similarly, AT&T, Verizon, and T-Mobile, with mobile behavior customer data in their arsenal, are positioned to swoop in as well on these mobile mobs.

Near-field communication (NFC) is far from being a standard in mobile payments, but the wave-and-pay technology may have gotten its biggest vote of confidence yet from the carrier community, as some forty-five of them have endorsed NFC and committed to support and launch services based on it. StrategyAnalytics (Strategyanalytics.com) forecasts that there will be sales of 1.5 billion handsets with SIM-based NFC capabilities worldwide between 2010 and 2016, and during that period there will be $50 billion in mobile transactions made. ABI Research (ABIResearch.com) believes that 85 percent of all new point-of-sale terminals shipped in 2016 will be NFC-enabled.

The forty-five carriers adopting the NFC standard include América Móvil, AT&T, AVEA, Axiata, AXIS, Bharti Airtel, Bouygues Telecom, China Mobile, China Unicom, CSL, Deutsche Telekom, Elisa Corporation, Emirates Integrated Telecommunications Company PJSC (du), Etisalat, Everything Everywhere, Globe Telecom, KPN, KT Corporation, Maxis, Mobily, MTS, Orange, Proximus, Qtel Group, Rogers Communications, Saudi Telecom Company (STC), SFR, SK Telecom, Smart, Softbank Mobile, Telecom Italia, Telecom New Zealand, Telecom Slovenije, Telefónica, Telekom Austria Group, Telenor, TeliaSonera, Telus, TMN, Turkcell, Verizon, VimpelCom, VIVA Bahrain, and Vodafone Group.

Start-ups such as Dwolla, Venmo (<https://venmo.com/>), and Square (Figure 4.11) are transforming mobiles into payment mechanisms at point-of-sale destinations. New geo-fencing technology is making it possible for apps loaded on

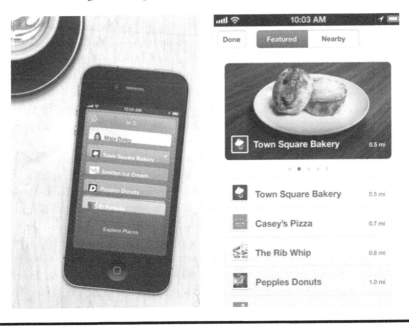

Figure 4.11   Square mobile pay systems.

mobiles to "communicate" with the payment systems of retail stores, thus eliminating the need to stand in line to make a purchase or even to interact with a sales clerk—taking away some market share from the $2 trillion credit card industry.

As we have seen, the mobile mobs are being driven by four American companies that have come to define twenty-first-century information technology, entertainment, and retailing: Amazon, Apple, Facebook, and Google are competing in these new mobile marketplaces with mobiles, apps, search, and social networking. Twitter is out because it does not have any hardware as of now. Only a couple of years ago, you could sum up what each did—Apple made hardware, Google was a search engine, Amazon was a bookstore, and Facebook was a social network—but mobile has change all of that.

All four now are pushing the mobile envelope beyond tech into communications, advertising, local marketing, retailing, entertainment, music, gaming, publishing, and cloud storage. Mobile is destroying the concept of stationary (PC) devices and replacing it with a shift toward powerful, full-fledged, hand-held computers. Two important facts are occurring because of this paradigm shift:

1. These mobiles are *intimately connected to individual users*. Unlike the family PC shared by those in a household, these devices define the behaviors, tastes, and preferences of the single human owner who carries them around in their pocket or purse.
2. These mobiles *generate a wealth of data* that comes to define the individual, which is why mobile analytics is the future of all of those areas these Fab Four firms are competing for. Each of these companies and their partners stand to benefit from this proliferation of mobile—a shift that will generate extraordinary gains in revenue, cash reserves, and market cap.

Amazon, Apple, Facebook, and Google are all going mobile and by doing so they are set to impact multiple industries, including the following market sectors:

■ *Mobile*: Apple has the iPhone and iPad, Google has the Motorola devices, Amazon has the Kindle Fire and is working on a Kindle mobile, and Facebook is working with HTC (<http://www.htc.com/us/>) on its device with extensive social network features integrated on a modified version of Android. Hardware companies impacted by the Fab Four are Acer, Asus, Dell, HP, Lenovo, LG, Microsoft, Nokia, Research In Motion, Samsung, and Toshiba.

■ *Entertainment*: YouTube and iTunes are roaring and growing, Google and Apple continue to work on Web-TV devices, while Amazon and Facebook have integrated Hulu and Netflix. Each company has launched a cloud-based, streaming music player. Games are the top apps on all of their properties. Companies impacted by the Fab Four are Comcast, DirectTV, DishNetwork, Disney, NewsCorp., TimeWarner, Viacom, EMI, Pandora, SonyMusic,

Spotify, WarnerMusicGroup, UniversalMusic, Activision, Blizzard, EA, Microsoft, Nintendo, Rovio, Sony, and Zynga.

■ *Media*: Apple is now a bookseller, while Amazon is now a publisher, Google is digitizing every book in the planet and all major magazines, and major publications like The Wall Street Journal now all have Facebook editions. Companies impacted are the New York Times, NewsCorp., Pearson, Random House, Scholastic, Viacom and Hachette.

■ *Tech*: All four will be offering some form of cloud computing storage services. Companies impacted are Box.net, Dropbox, EMC, Rackspace, Salesforce. com, and VMWare.

■ *Advertising*: All four have algorithm-driven—international and local ad networks—using mobile search, social media, and location- or interests-based behavioral analytics. Companies impacted are Interpublic Group, Omnicom Group, Publicis Groupe, Starcom, Foursquare, Groupon, OpenTable, and SuperMedia.

■ *Retail*: All four are competing for a share of the $140 billion local ad market with aggressive, time- and location-specific discount offers; they all have mobile online checkouts, and tap-and-pay wallets. Companies impacted are BestBuy, eBay, RadioShack, Target, Walmart, American Express, MasterCard, and Visa.

All four players are well aware that these mobile mobs can be triangulated via the parameters of time, interests, location, and anonymous human and device demographics, such as digital fingerprints and other mobile tracking mechanisms. Enterprises, retailers, brands, and marketers need to focus on mobile sites, customized apps, and mobile analytics. They must reach the consumers, consistent with how these moving mobiles are used—on-the-fly—via quick bursts of interactive communications with ads, deals, contests, games, coupons, offers, etc.

Chapter 5

Mobile Analytics

5.1 Mining Mobiles

For years, mobiles have been getting smarter and more ubiquitous. The momentum in marketing technologies is clearly now with mobiles that can be carried around, broadcasting their interests and locations. More importantly for mobile developers, brands, and marketers, sites and apps have turned mobiles into great data aggregators and delivery platforms for triangulated modeling and real-time advertising. The data mining of mobiles involves knowledge discovery from the nontrivial process of identifying valid, novel, and potentially understandable patterns and relationships in mobile behaviors.

The objectives of data mining mobile devices are twofold: *prediction* and *description*—prediction of unknown or future values of selected variables, such as interests or location of mobiles, and description in terms of human behavior patterns. Description involves gaining "insights" into mobile behaviors, whereas prediction involves improving decision making for brands, marketers, and enterprises. This can include the modeling of sales, profits, effectiveness of marketing efforts, and the popularity of apps and a mobile site. The key is to realize the data that is being aggregated and how to not only create and issue metrics on mobile activity, but more importantly, how to leverage it via the data mining of mobile devices to improve sales and revenue.

For years, retailers have been testing new marketing and media campaigns, new pricing promotions, and the merchandising of new products with freebies and half-price deals, as well as a combination of all of these offers, in order to improve sales and revenue. With mobiles, it has become increasingly easy to generate the data and metrics for mining and precisely calibrating consumer behaviors. Human habits are intractable, inexplicable, and ingrained—they are so strong, in fact, that they cause consumer brains to cling to them at the exclusion of all else.

Today, those mobiles in people's pockets or purses are the bright beacons broadcasting what consumers want, when and where they want it. Brands and companies leveraging mobile analytics can be more adept at identifying, co-opting, and shaping consumer behavior patterns to increase profits. Brands and mobile marketers that figure out how to induce new habits can enhance their bottom lines. Inducing a new habit loop can be used to introduce new products, services, and content via the offer of coupons or deals based on the location of mobiles.

Billions of people around the world use the Internet daily with mobiles, creating compelling opportunities for businesses to advertise their products and services to customers—no matter where they are. However, there are considerable technological challenges for advertisers seeking to engage a mobile audience. These include mobile fragmentation, post-click Web engagement, and consumer usage behavior. Advertising platforms for mobile marketing, such as MobiVite.Net (<http://www.mobivite.net/welcome>), have introduced programs to help interactive agencies and mobile ad networks overcome the technological barriers of their clients.

The worldwide CRM (Customer Relationship Management) applications market is $18.2 billion; the mobile advertising market is expected to be worth billions of dollars according to analysts' reports indicating AdMob, Millennial Media, and Apple's iAd market share. Several studies indicate that mobile marketing budgets will increase considerably and so will consumers' expectations of how they interact with businesses of their choice.

However, due to unique technology, integration, and consumer experience challenges posed by mobile, many brand advertisers and small- to mid-sized businesses (SMBs) are struggling to integrate mobile into their marketing campaigns in a meaningful way. This is an excellent opportunity for interactive agencies and mobile ad networks to deliver highly relevant advertising and marketing campaigns, end-to-end.

Mobile ad spending in the United States is projected to rise to 80 percent this year, to $2.61 billion, according to AdAge.com, in their new report that estimates that mobile ad spending will reach 1.45 billion, up from $769.6 million, and that Google is getting more than half of the pie. Facebook could have the upper hand in comparison to Google and Apple, among others who have already launched mobile ad campaigns.

Facebook knows it users and gathers their information in the form of "likes" so the social media giant will know what types of advertisements to use in targeting individuals. With so many different platforms—from iPhones to tablets to Android's multitudes of models—it is difficult to build an ad that can be guaranteed effective. However, new standardizations are becoming popular, and liquidity through real-time bidding and consolidation will help address these problems and bring mobile ad spending of agencies into the market at scale for the first time.

For example, Flurry Analytics offers a free analytics tool for mobile app developers; it now also supports HTML5. The tool can already be incorporated into apps for the main mobile phone operating systems of Apple, Android, BlackBerry, and

Windows mobiles. This analytics tool can gather anonymous, aggregated usage and performance data, and is already in use by 150,000 apps and 60,000 companies. The addition of support for the new HTML5 standard means Flurry can be used to track rich content running in a browser on any mobile, rather than just working with dedicated apps that are made specifically for each type of device.

5.2 Mining Mobile Sites

For starters, the analysis of the mobile site experience needs to take place; this includes metrics rarely including the quality of the mobile site experience, a distinct concept and according to a recent report from Forrester (Forrester.com), that is a big mistake: Without dedicated customer experience metrics, brands, marketers, and companies cannot tell whether the mobile site experience actually got better or how changes in the quality of that experience affected the site's business performance. Here is a summary of some practices that will start to fill this gap. According to the market research company, Forrester, at a minimum, Web customer experience (WEM) professionals should track what mobile customers thought of the following factors:

- The overall mobile visit experience and satisfaction are the most common metric. Ways to determine include surveys that are signal or emotion based.
- Visitors tend to focus on three basic things when evaluating a mobile site: usefulness, ease-of-use, and how enjoyable it is. Metrics should measure these criteria with completion rates and survey questions.
- Brands and companies should use observational research to figure out what else their customers need to deem a site visit "good" and include at least one metric for each of those criteria. For example, visitors to a financial services site might expect it to be secure and up-to-date. In that case, customer experience pros would need a way to measure whether or not visitors felt as though the site lived up to those descriptions.

This year (2012), mobile will come into its own as a mainstream marketing medium, as more businesses allocate larger budgets to mobile advertising and, especially, to developing mobile sites. And that is a good thing, considering that usage of the mobile Web is exploding. In fact, a Google report (Google.com) shows that the percentage of search queries stemming from mobile continues to grow for many industries:

- Restaurants: 29.6 percent
- Auto: 16.8 percent
- Electronics: 15.5 percent
- Finance/Insurance: 15.4 percent
- Beauty/Personal: 14.9 percent

And those numbers are just the tip of the iceberg. Mobile site analytics can help the brand and companies solve the mystery of how mobile consumers are engaging and interacting with their site. The first rule of thumb is do not assume a traditional site renders well on smart screens. Tools like MobileMoxie (<http://www.mobile-moxie.com/>) reveal how a mobile site appears via a handheld device—determine if there is a need of a mobile facelift; most likely, this will be the case. Before making any changes, cross-check the traditional site's analytics to learn about the number of mobile visitors; in doing so, the developer will be able to better target their mobile site design, content, and optimization efforts for specific mobile experiences.

Visits/new visits: If the brand or company notices that their overall traffic has a high percentage of mobile visitors, then they will be able to justify developing a mobile-friendly site much more quickly. If the percentage of mobile visitors is over 20 percent, that will be a clear indication that a mobile site is needed. With higher mobile traffic, of say over 40 percent of mobile searches, most will be local; this is critical because over 61 percent of consumers follow up with some immediate action such as phone calls, and 58 percent with in-store visits, according to studies by Google (Google.com).

With that in mind, design the mobile site by putting basic business and contact information up front, as well as providing click-to-call and mapping and navigation information. It is also noteworthy to watch visits increase as mobile penetration and mobile efforts improve with call-to-action prompts.

Pages per visit: How deep are mobile users navigating within the site? Due to small screen sizes and the sensitive touch-screens, mobile users do not click as much as desktop users. Also, desktop users typically visited an average of five pages, whereas mobile visitors only viewed up to two. Because of this, a mobile site should place its product and service descriptions succinctly and prominently to eliminate endless scrolling and clicking. If clicking-through is necessary, ensure links, navigation, and buttons can be accessed and are large.

Bounce rate/average time on site: Some 80 percent of consumers immediately quit their shopping if they have a bad experience on mobile devices. Average time and bounce rate can be indicators that a mobile site is not performing well on mobiles or living up to visitor expectations, from both a content and functionality standpoint.

The following is a short analysis of how mobile Web analytics are implemented across a few top sites that were picked from the Alexa (Alexa.com) rankings:

- **Amazon** uses inline JavaScript to build the URL of an analytics beacon. The code is minified and the total size, including script, noscript, img, and a div tag is 5 KB.
- **EBay** and **Walmart** do not seem to have any form of analytics code sent on the client side, probably some form of tracking is done on the server, it being transparent to the client.
- **Best Buy** directly embeds an Omniture beacon in the page. There is no extra JavaScript to download and execute; it also has an embedded Atlas beacon image.

- **Groupon** has separate sites for enhanced and featured devices. They both use rum.js, a 3 KB script that generates a beacon URL for Google Analytics.
- **IKEA** has its analytics code bundled inside a 50 KB, zipped JavaScript file. The size of the analytics code is 67 KB minified, which becomes 22 KB gzipped.
- **Target** uses Omniture for tracking and includes a 47 KB (17 KB gzipped) JavaScript in the page. An interesting fact is that their analytics script is more than twice as large as their JavaScript, which is only .5 KB zipped.
- **Dell** also uses Omniture and downloads a 50 KB (19 KB gzipped) JavaScript.
- **IBM** sends a beacon to unica.com. The beacon URL is built by a downloaded JavaScript file. The file size is 18.6 KB minified (6.8 KB gzipped).
- **Microsoft** uses a directly embedded beacon image that hits webtrends.com.
- **Apple**, **SAP**, **Canon**, **Asus**, **Toshiba**, **Fujitsu,** and **Acer** do not serve optimized content for mobiles.

5.3 The Metrics of Mobile Analytics

The following are some key data mining methodologies and objectives of the data mining of mobile devices process. Now, they may come from using app or site analytics or some deep machine learning processes, but they all come to the same objectives:

1. *Pattern*: An expression describing facts about a mobile dataset.
2. *Relation*: The description of relations between mobile data attributes or patterns.
3. *Process*: A multistep procedure involving data preparation, cleaning, mining, etc.
4. *Novel*: Discovering a not-yet-known knowledge discovery about mobile behaviors.
5. *Understandable*: Knowledge that brands and marketers can grasp and use.

The following are some key steps in data mining and knowledge discovery of mobiles:

1. Develop an understanding of the mobile dataset and goal of the inductive and deductive data analysis.
2. Create a target dataset or subset.
3. Preprocess the data by cleaning it: remove outliers, handle missing values.
4. Transform the data via dimensionality reduction to find useful features.
5. Select the data mining tasks: clustering, text mining, classification.
6. Select a data mining algorithm: self-organizing map (SOM), decision tree.
7. Perform the mobile mining analysis.
8. Interpret the discovered mined patterns, or hidden relationships.
9. Consolidate the discovered knowledge.

Keep in mind that the primary task of data mining is prediction and description, which commonly involves fitting models to or determining patterns from mobile device behavioral data. This is commonly known as the knowledge data discovery (KDD) process (Figure 5.1).

The traditional data mining process involves several steps, some of which involve a wide array of inductive and deductive techniques and software. Data mining of mobile devices also involves a deep understanding of how mobile data is created and captured by both apps and mobile sites. It also requires knowing how to prepare the data and what techniques to use for the given tasks, how to use the tools and techniques, how to validate the results, and most importantly, how to deploy the results in order to improve the business process and increase sales and revenue or improve a brand and customer loyalty.

Keep in mind that data mining has been around for years, and the software tools are highly robust and mature. For information on the tools and techniques, go to KDnuggets.com, a site devoted to data mining, big data, and analytical software, jobs, consulting, courses, white papers, and more.

5.4 Google Mobile Analytics

The Web is inexorably moving in the direction of mobile browsing. As more people purchase mobiles and are able to access the Web from anywhere, knowing what pages they want to see in a mobile-friendly format is essential. Google has integrated mobile tracking directly into its new analytics platform, and the data is indispensable.

Google analytics (Figure 5.2) makes the mobile site information easily accessible through either applying a secondary search dimension on a per-device basis, or as a total percentage of the total traffic through advanced segments. Again, having sophisticated custom segments to accurately track keywords can show a brand or company where and how visitors are finding a mobile site and through what keywords.

The tool also reports on what operating systems visitors are running, such as Apple or Android, and what mobile models are landing at the site. Once again, aside from the obvious data mining capability, client management is hugely affected by this new integration. The analytical tool allows brands and companies to see where the mobile users are going and how long they are staying.

Right off the bat, the tool demonstrates two things that are extremely interesting. As a percentage of the total traffic to the site, 6.11 percent of all traffic is from a mobile platform; and if the site developers click on the sub-category devices, they can see which specific devices are visiting the site. Some 6 percent of total site traffic is substantial enough to warrant an investigation as to where these visitors are going, and to consider whether these pages should have a mobile page built for them.

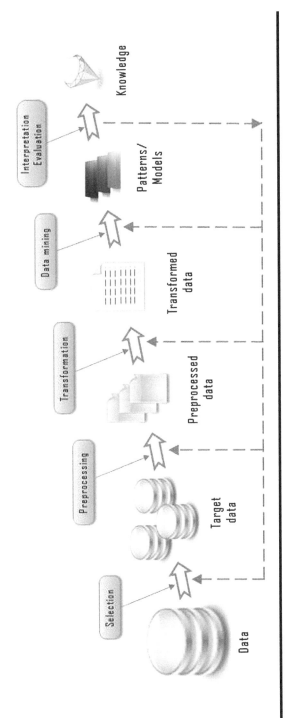

Figure 5.1   The sequential KDD process.

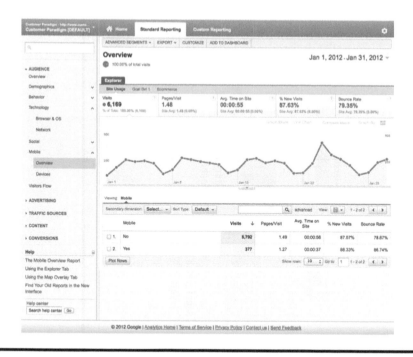

Figure 5.2 Google analytic overview of mobile activity.

5.5 An Artificial Intelligence (AI) App

Mobile analytics is deeply rooted in artificial intelligence (AI) techniques and technologies, such as clustering SOMs and decision trees. For example, Apple's voice recognition Siri is built on AI. Now, there is another application based on AI; her name is Evi (<http://www.evi.com/>). Her answers are quite different from those of Siri, with a different kind of intelligence and accent. Evi is a next-generation AI now being launched via her own "conversational search" mobile app.

Evi uses natural language processing and semantic search technology to infer the intent of the user's question; she then gathers information from multiple sources, analyzes them, and returns the most pertinent answer. She does this by going beyond word matching and instead reviews and compares facts to derive new information for the user.

The user can state, "I need a coffee" and she will tell them what coffee shops are nearby, along with addresses and contact details. Evi understands what the user means and gives them only the information they really need. Evi was designed to draw upon what is now approaching 1 billion facts contained in their "True Knowledge" database. In addition to the database, Evi's pool of knowledge is extended through connections with popular APIs (application programming interfaces), such as Yelp. com; for example, when asking for a nearby Chinese restaurant, Evi will also help the user make the reservations within the app itself (Figure 5.3).

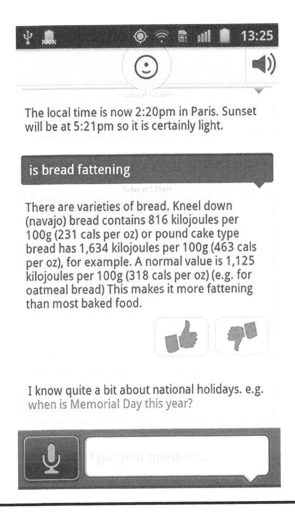

Figure 5.3   Evi has a clean interface.

Compared to Siri, which is primarily used for voice-activated phone control with additional search capabilities, Evi is designed to "take the search out of searching" by leveraging True Knowledge proprietary AI technology to deliver filtered intelligence in a concise and relevant format. Evi can be downloaded for free in the Android Marketplace and in the iTunes App Store, where a U.S.$0.99 cost covers language processing license fees.

5.6 Clustering Mobiles

One of the first tasks for data mining devices is that of clustering, which is using SOM software to partition a mobile dataset into subsets of "similar" data autonomously, automatically, without bias or prior knowledge by the analyst about the properties or existence of these subsets. This type of data mining analysis is also known as *unsupervised learning* because the SOM program partitions the dataset into clusters of similar subsets. For example, a clustering analysis might autonomously discover similarities in ways visitors behave upon arrival at a landing page of a mobile site. Or, using an SOM program, the analyst might discover different clusters of users who have downloaded and installed an app.

Clustering is the partition of a dataset into subsets of "similar" data, without using a *priori* knowledge about properties or existence of these subsets. For example, a clustering analysis of mobile site visitors might discover a high propensity for Android devices to make higher amounts of purchases of, say, Apple mobiles. Clusters can be mutually exclusive (disjunct) or overlapping. Clustering can lead to the autonomous discovery of typical customer profiles.

Clustering detection is the creation of models that find mobile behaviors that are similar to each other; these clumps of similarity can be discovered using SOM software to find previously unknown patterns in mobile datasets (Figure 5.4). Unlike classification software, which analyzes for predicting mobile behaviors, clustering is different in that the software is "let loose" on the data; there are no targeted variables. Instead, it is about exploratory autonomous knowledge discovery. The

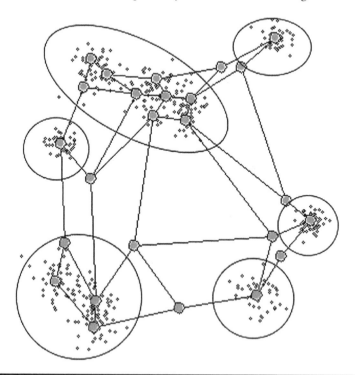

Figure 5.4   Clusters discovered by a SOM.

clustering software automatically organizes itself around the data with the objective of discovering some meaningful hidden structures and patterns of mobile behaviors. This type of clustering can be done to discover key words or mobile consumer clusters, and it is a useful first step for mining mobiles. It allows for the mapping of mobiles into distinct clusters of groups without any human bias.

What distinguishes clustering from classification, which we discuss in Section 5.9, is that the mining does not rely on a predefined class of, say, buyers versus non-buyers. This type of clustering analysis does not start with the objective of prediction. Instead, the datasets are grouped together on the basis of self-similarity or unique clusters. Clustering is often performed as a prelude to the use of classification analysis using rule-generating or decision-tree software for modeling mobile device behaviors.

One common use for this type of undirected knowledge discovery is "market basket analysis," which seeks to answer "what items sell together" or "why do some mobiles behave the same way?" Once the data has been broken into key segments or clusters, the analyst or marketer can begin the process of discovering interesting mobile behavior patterns in various subgroups. Following are some of the key steps for conducting a clustering analysis:

1. Identify the mobile behaviors of most interest from app usage to mobile site traffic.
2. Compose a clustering model using SOM software to autonomously discover data subsets.
3. Evaluate the significant finding of the clusters, such as the number of samples and their median ranges.
4. Use text mining and classification software to further discover key "features" of the autonomous clusters.

Undirected knowledge discovery is a good initial step for mobile miners to take by mobile miners, analysts, developers, and marketers to generate ideas that can be verified by supervised learning methods. A market basket analysis can lead to questions about *why* certain products sell at the same time, or *who* is buying a particular combination of products or services, or *when* purchases tend to be made. A clustering analysis can be used to discover what mobiles respond to certain offers, deals, or coupons; and because it is self-organized, no human bias is introduced into the process of knowledge discovery.

As previously mentioned, this type of undirected, unsupervised type of knowledge discovery is best performed using neural network SOM software. One such program is Viscovery SOMine® (<http://www.viscovery.net/somine/>), which is software for explorative data mining, visual cluster analysis, statistical profiling, and segmentation. This kind of software can be used to perform market basket analyses of mobiles; to discover who buys what, when, and where based on their location or interests; and to ascertain the behaviors and features of mobiles.

Market basket analysis using a SOM is useful in situations where the marketer or brand wants to know what items or mobile behaviors occur together or in a particular sequence or pattern. The results are informative and actionable because they can lead to the organization of offers, coupons, discounts, and the offering of new products or services that prior to the analysis were unknown.

Clustering analyses can lead to answers to such questions as *why* do products or services sell together, or *who* is buying what combinations of products or services; they can also map *what* purchases are made and *when*. Unsupervised knowledge discovery occurs when one cluster is compared to another and new insight is revealed as to why. For example, SOM software can be used to discover clusters of locations, interests, models, operating systems, mobile site visitors, and app downloads, thus enabling a marketer or developer to discover unique features of different consumer mobile groupings.

5.7 Mining Mobile Mail

comScore® (<http://www.comscore.com/>) recently released its "2012 U.S. Digital Future in Focus" report (<http://www.comscore.com/Press_Events/Presentations_Whitepapers/2012/2012_US_Digital_Future_in_Focus>) and it reported on a startling trend in mobile email. One of the more telling stats concerns email use among those in their teens and twenties. According to the report, Web-based email use among 12- to 17-year-olds dropped 31 percent in the past year, while use among those 18 to 24 years old saw an even bigger drop of 34 percent. Some of that can no doubt be attributed to Facebook and other email alternatives, but a big factor is the growth of email use on mobile devices; both of those age groups saw double-digit growth in that respect, with mobile email use jumping 32 percent among 18- to 24-year-olds.

In terms of sheer growth in the past couple of years, however, there is not much that matches the trajectory of tablets. comScore notes that U.S. tablet sales over the past two years have topped 40 million, a figure that it took mobiles as a category a full seven years to reach. Another area that saw some considerable growth is in digital downloads and subscriptions, such as e-books, which jumped 26 percent compared to the previous year, leading all other areas of e-commerce. All of these trends suggest that mobile miners and marketers need to strategically plan for future analyses.

5.8 Text Mining Mobiles

Another technology that can be use for data mining mobile devices is text mining, which refers to the process of deriving, extracting, and organizing high-quality information from unstructured content, such as texts, emails, documents, messages, comments, etc. Text mining means extracting meaning from social media and customer comments about a brand or company in mobile sites and app reviews.

This is a different variation of clustering programs; text mining software is commonly used to sort through *unstructured content* that can reside in millions of emails, chat, Web forums, texts, tweets, blogs, etc., that daily and continuously accumulate in mobile sites and mobile servers. Text analytics generally includes such tasks as (1) the categorization of taxonomies, (2) the clustering of concepts, (3) entity and information extraction, (4) sentiment analysis, (5) summarization, as well as (6) the autonomous creation of ontologies.

Text mining (Figure 5.5) is important to the data mining of mobile devices because, increasingly, companies, networks, mobile sites, enterprises, and app servers are accumulating a large percentage of their data in unstructured formats, which is impossible to analyze and categorize manually. Text mining refers to the process of deriving an understanding from unstructured content through the division of clustering patterns and the extraction of categories or mobile trends using machine learning algorithms for the organization of key concepts from unstructured content.

Text mining usually involves the process of structuring the input text, usually by parsing it, along with the addition of some derived linguistic features and the removal of others, and subsequent insertion into a structured format such as a database and deriving these clusters and patterns for evaluation and interpretation of

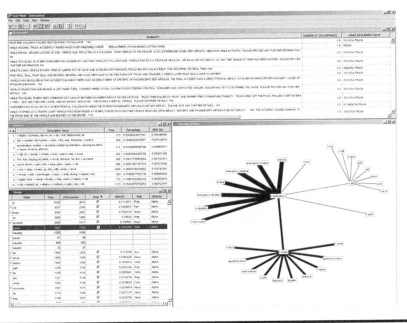

Figure 5.5 The results of a text mining analysis.

their output. Text mining usually refers to some combination of relevance, novelty, and interestingness. Typical text mining tasks include the categorization and clustering of unstructured content and concept, with the final products being some sort of granular taxonomy, sentiment analysis, document summarization, and entity relation modeling.

Text mining can be used to gain new insight into unstructured content from multiple data sources, such as a social network of a mobile site or an app platform. Text analytical tools can convert unstructured content and parse it over to a structure format that is amenable to data mining of mobile devices via classification software. For example, all the daily emails or visits that a mobile site accumulates on a daily basis can be organized into several groupings, such as those mobiles seeking information, service assistance, or those complaining about specific products, services, or brands. Text mining can also be used to gauge sentiment regarding a brand or company.

Text analytical software allows analysts and marketers to narrow down a set of emails, texts, tweets, blogs, or other mobile generated content and autonomously organize them into multiple categories, such as *awesome, well done, great, good,* or *OK.* Text mining involves applying computationally intensive algorithms to large collections of unstructured content. Aside from organizing key concepts, mobiles can also be organized by text mining software into key segments, along service and product lines. For example, mobile site visitors or app downloads can be grouped into those that are interested in sports, movies, books, games, financials, music, or other topical categories.

Text analytics can also discover key metrics for the mobile miner, developer, analyst, or marketer and combine them into an intuitive visualization about brand buzz, brand sentiment, and passion intensity. The autonomous clustering of words from emails, instant messages, chats, texts, tweets, and other information from mobiles can be used for the creation of matrices of words along key consumer categories. The true value of text mining is that it can speed up the analysis of unstructured content into clear and concise categories for data mining mobile devices (Figure 5.6).

Mobile marketers, developers, and brands need to consider how to incorporate time, demographics, location, interests, and other mobile available variables into their analytics models. Clustering, text, and classification software can be used to accomplish this for various marketing and brand goals. Clustering software analyses can be used to discover and monetize mobile mobs. Text software analyses can discover important brand value and sentiment information being bantered about in social networks. Finally, classification software can pinpoint important attributes about profitable and loyal mobiles. Classification often involves the use of rule-generating decision-tree programs for the segmentation of mobile data behaviors.

Figure 5.6   Data mining mobile devices.

Case Study: Placecast

Placecast   (<http://placecast.net/shopalerts/payments.html>)
works with many brands via their ShopAlerts® service, which
is an opt-in, location-based mobile marketing solution that
delivers highly relevant SMS messages customized for each
customer based on their location, time, user preferences, and
CRM data; the following are some of their clients:

- The North Face: Summit Signals, the North Face location-
  based messaging program, enables fans of the brand to
  stay connected about the latest gear, local events, sponsored sports events, and tips for outdoor activities.
- White House Black Market: Fashion Alerts, the location-
  based mobile program from White House Black Market,
  offers the brand's customers exclusive offers and promotions as well as insider scoop on new product arrivals and
  VIP store events.

- Sonic: SONIC® Drive-In customers participated in the restaurant's Sonic Signals programs and received special promotions and offers when they enter geo-fences-created areas around SONIC restaurants.
- O2: Placecast collaboration with O2, the second-largest mobile operator in the United Kingdom, enables ShopAlerts to be utilized directly by the carrier for more than twenty brands and advertisers and more than one million opted-in subscribers.

5.9 Classifying Mobiles

There are two major objectives to classification via the data mining of mobile devices: *description* and *prediction*. Description is an understanding of a pattern of mobiles behaviors and to gain insight—for example, what devices are the most profitable to a mobile site and app developer. Prediction, on the other hand, is the creation of models to support, improve, and automate decision making, such as what highly profitable mobiles to target in an ad marketing campaign via a mobile site or app. Both description and prediction can be accomplished using classification software, such as rule-generator and decision-tree programs. This type of data mining analysis is also known as *supervised learning*.

The description and prediction of mobile behaviors is made possible by data mining machine learning-based software. Machine learning is a branch of AI; it is a discipline that allows software to make predictions based on empirical mobile behaviors. These mobile behaviors can be aggregated from a mobile site or downloaded apps that contain usage information from visitors or app users. In the classification of mobiles, historical triangulated server data is used to construct predictive models, which can be used to improve consumer loyalty and increased sales and revenue for a mobile site or app developer.

For example, a mobile analyst or marketer can take advantage of segmenting the key characteristics of mobile behaviors over time to discover hidden trends and patterns of purchasing behaviors. Machine learning technology can discover the core features of mobiles by automatically learning to recognize complex patterns and make intelligent decisions based on mobile data, such as what, when, where, and why certain mobiles have a propensity to make a purchase or download an app, while others do not.

The strategy for the analyst or marketer is to gather and model consumer mobile data to make marketing to these devices more personal, relevant, and comprehensive. This requires capturing, analyzing, and acting on mobile actions and leveraging from these behaviors to entice the downloading of an app, the adjustment of

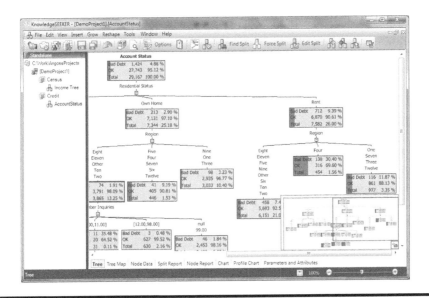

Figure 5.7   An example of decision tree segmentation ability.

prices for products or services, the offer of discounts—the applications are many (Figure 5.7).

Classifying mobiles enables the positioning of the right product, service, or content to these moving devices via precise messages on a mobile site, or the targeting of an email, text, or the creation of key features to an app. The success of mining mobiles involves strategic planning and measured improvement of multiple, predictive, evolving models.

The marketer or developer will need to use classification software known as rule-generators or decision-tree programs. A good source for downloading demos and actual programs on the Internet, as previously mentioned, is KDnuggets.

This type of software is commonly based on a proven class of techniques that includes classification and regression trees (CARTs), C4.5, and chi-squared automatic induction (CHAID) algorithms, which have been in the marketplace for years. Machine learning software allows developers and marketers to perform mobiles segmentation, classification, and prediction; these types of programs can generate graphical decision trees as well as predictive rules for classifying mobile behaviors.

Decision trees are powerful classification and segmentation programs that use a tree-like graph of decisions and their possible consequences. Decision-tree programs provide a descriptive means of calculating conditional probabilities. Trained with historical data samples, these classification programs can be used to predict future mobile behaviors. They divide the records in a "training" dataset into disjoint subsets, each of which is described by a simple rule on one or more mobile data fields; for example,

IF	Model: HTC EVO 4G LTE
AND	OS: *Android* 4.1, Jelly Bean
AND	Geolocation: El Paso, TX ZIP 79902
THEN	Offer: Torta coupon

A decision tree takes as input an objective, such as what type of app to offer, described by a set of properties from historical mobile behaviors or conditions, such as geo-location, operating system, and device model. These mobile features can then be used to make a prediction, such as what type of app to offer to a specific mobile. The prediction can also be a continuous value, such as total expected coupon sales, or what price to offer for an app.

When a developer or marketer needs to make a decision based on several consumer factors, such as their location, device being used, total log-in time, etc., a decision tree can help identify which factors to consider and how each factor has historically been associated with different outcomes of that decision—such as what products or services certain mobiles are likely to purchase based on observed behavioral patterns over time (Figure 5.8).

A key technology is decision-tree software, which is a technique that recursively splits and divides data space into sub-regions with decision hyper-planes orthogonal to coordinate axes. For example, a decision tree can analyze a database of fruit and ask such questions as color, size, and shape; it can determine which is the most important attribute for classifying, say, an apple or banana. By measuring the value of each property (e.g., color (green)), it can prioritize what attribute to use for the classification of these different fruits. Color would probably be the most information intensive, followed by size and shape. It is through this machine learning interrogative process that decision trees can be used to classify mobile behaviors and profiles.

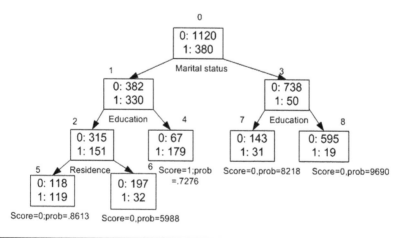

Figure 5.8 A decision tree can be used to segment mobile consumer features.

Modern classification decision-tree tools are highly intuitive; their interfaces are easy to navigate and use, their segmentation results from predictive rules, and graphical trees can readily be inspected for quick insight by mobile marketers and brands. The value of these tools is that the graphs and rules can easily be understood and applied toward mobile marketing objectives. For example, predicting if a mobile will make a purchase, a decision tree can examine multiple factors such as the number of mobile site visits, type of payment, operating system of the device, products viewed, or the mobile's triangulated location.

Classification decision trees go through thousands of iterations in which independent attributes, such as mobile location, time of day, mobile model, etc., representing mobile behaviors and features are measured in terms of their "information gain." For mobile marketers and developers, they automate the identification of key device variables as they relate to sales, potential future revenue, and consumer loyalty. Almost always, decision trees produce compact models to identify the most valuable attributes for predicting consumer behaviors with only a few conditional rules; they are a form of "information compression" for mobile marketers.

One common advantage of using decision trees is to eliminate a high number of noisy and ineffective consumer attributes for predicting, say, "high customer loyalty" or "likely to buy" models. Developers and marketers can start with hundreds of mobile attributes from multiple data sources and, through the use of decision trees, they can eliminate many of them in order to focus simply on those with the highest information gain as they pertain to predicting high loyalty or potential revenue growth from mobile features and behaviors.

These classification predictive tools are both easy to use and understand, as well as computationally inexpensive for predicting device behaviors. There is also software-as-a-service (SaaS) from Zementis (<http://www.zementis.com/>), which offers an on-demand predictive analytics decision engine hosted on a cloud. A framework for accomplishing and leveraging these predictive models should be flexible and ongoing as conditions change.

Case Study: BayesiaLab

BayesiaLab offers Bayesian classification algorithms to automatically cluster mobiles behavioral variables into distinct consumer groups. A Bayesian network is *a graphic probabilistic model* through which a user can acquire, capitalize on, and exploit knowledge. Particularly suited to *taking uncertainty into consideration*, a Bayesian network or belief network is a probabilistic graphical model that represents a set of random variables and their conditional dependencies.

For example, a Bayesian network could represent the probabilistic relationships between certain mobiles and consumer preferences. Given these preferences, a network can be used to compute the probabilities of the presence of Apple or Android mobiles and certain products or services being offered by a retailer or brand. A Bayesian network is used to analyze data in order to diagnose clusters and the resulting probability distribution of these effects. A Bayesian network is a graphical model that encodes probabilistic relationships among variables of interest and has several advantages for the discovery of hidden relationships for data mining mobile devices.

There are several advantages to the use of Bayesian network software. First, because the model encodes dependencies among all variables, it readily handles a situation where some data entries are missing, which is a common dilemma in mining mobile retail data. Second, a Bayesian network can be used to learn causal relationships and hence can be used to gain understanding about a problem domain and to predict the consequences of intervention—for example, what offers or deals to make to what mobiles. Finally, because the model has both causal and probabilistic semantics, it is an ideal representation for combining prior knowledge in large datasets for mining mobiles; in other words, the analyst or marketer knows bests how to interpret the finding of a Bayesian network (Figure 5.9).

The software is extensive, no cause is overlooked, and all possibilities are considered. BayesiaLab is an ideal tool for analyzing the uses and attitudes of a group of customers, satisfaction questionnaire analysis, brand image analysis, segmentation, clusters and groupings of customers or products, and assessment of appetence scores in relation to new products and multiple mobiles. The BayesiaLab market simulator is a program through which a mobile developer or marketer can compare the influence of a set of competing offers in relation to a defined population of mobiles, consumer behaviors, and groupings.

The databases relative to the mobile customers can be used to elaborate profiles using data mining methods. These profiles can then bring objective information to the marketing department; they can also be used to reduce the cost of mobile ad campaigns by selecting only the prospects that have a high probability to reply positively. For example, BayesiaLab was used for the profiling of customers with respect to a bank product that has various modalities. The database used contains variables describing the customer, such as age, socio-professional, gender, etc.; it also had bank account variables such

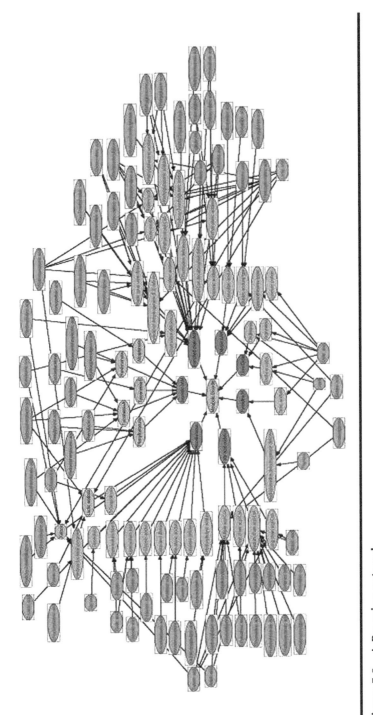

Figure 5.9   A Bayesian network.

as facilities, rate of consumption, etc., and the modality of its bank products.

Bayesian networks represent all the probabilistic relations that hold in a dataset; for example, their Markov Blanket Learning algorithm allows focusing a search only toward the variables that really characterize the target variable. The Bayesian network can be used to predict the bank product for new customers. As BayesiaLab returns the probability associated to that prediction, it is possible to use this probability to reduce the cost of the campaigns by using this probability to select the prospects.

5.10 Streaming Analytics

Mobile developers and marketers can also enlist both deductive and inductive "streaming analytical" software, which is "event-driven," to link, monitor, and analyze mobile behaviors. These relatively new streaming analytical software products react to "device events" in real-time. There are two main types of streaming analytical products; there are *deductive* streaming programs that operate by user-defined business rules and are used to monitor multiple streams of data, reacting to device events as they take place. For example, when a mobile enters a Wi-Fi location, or a text is sent, triggering a marketing counteraction, such as an offer, a discount, a coupon offer, an alert, an invitation, etc., based on business rules created by a marketer, a brand, or an enterprise.

There are also *inductive* streaming software products that use predictive rules derived from the data itself via clustering and decision-tree algorithms. These inductive streaming products build their rules from global models involving the segmentation and analysis from multiple and distributed data clouds and networks of device behaviors. These deductive and inductive software products can work with different data formats, from different locations, to make real-time predictions using multiple models from massive digital data streams.

These types of streaming analytic software products support the processes of analyzing an assortment of behaviors, locations, interests, demographics, lifestyles, geo-spatial, operational, etc., and other information for the personalization of offers to mobiles. Real-time streaming becomes a continuing and iterative process in which marketing decisions and actions are incessantly refined and perfected over time.

Everything can be measured in terms of revenue, loyalty, relevancy, satisfaction, speed, and performance. Every single metric resides in some digital format amenable to refinement and improvement. In the end, the data mining of mobile devices is about delivering targeted and relevant content and offers, for customer retention and loyalty, as well as up- and cross-selling.

It is important that this classification of mobile behaviors be based on analyses tested and measured in terms of total sales or other metrics that the mobile marketer and developer determines are of most value for sustained growth and revenue. The advantages of static and streaming behavioral analytics for mining mobiles are many; here are just a few:

1. Perform up-selling and cross-selling of multiple products and services.
2. Reduce uncertainty, predict with precision, and optimize performance.
3. Gain insight into, model, and monetize mobile behavioral patterns.
4. Provide consumer personalization while preserving privacy.
5. Achieve higher profits via improved consumer marketing.
6. Improve efficiency at every level of operations.
7. Spot new growth opportunities and markets.
8. Increase customer satisfaction and loyalty.

Data mining of mobile devices enables the discovery of their preferences and needs directly from their behaviors and purchases. These cyclical feed-forward interactions can provide vital business intelligence that traditional marketing techniques cannot match in both accuracy and speed. Today, companies and brands can subscribe or construct their own behavioral analytical systems in support of data mining mobile devices, enabling them to process their "events" as they happen in order to respond appropriately—much like the neighborhood merchant of yesteryear who remembered his customers' tastes and took pains to please them in order to retain them for life.

Enterprises, brands, and marketers do not want to offer consumers products or services they do not want, or content that is not relevant to them; this is not only wasteful, but also intrusive. Instead, data mining mobiles devices is about using clustering, text, and classification software that can provide the right product to the right mobile at the right time and place. Care must be taken to protect the consumer's privacy and security. Mobile behaviors are the most valuable assets marketers and their clients have, and they need to protect them and not share them with others.

5.11 Mining Mobile Mobs

Mobile sites provide a gold mine of device data, everything from browsing behavior patterns to demographics, transactional histories, sources of traffic, the effectiveness of search and social marketing, number of downloads, keyword used, and cross-selling propensities. With every "event," mobiles are communicating with companies and brands at their mobile sites as to their needs and desires. With mobile mining, developers and marketers can leverage these events, most of which start at their site but cascade across other operational systems within enterprises, many of which are increasingly mobile, such as a brand or an enterprise app.

In an ever-increasingly mobile networked environment, advocates of brands enable marketers and enterprises to build customer loyalty by providing them with a social platform by which they can communicate with their friends about what they *like* about a product or service. These existing and new friends can promote digital word-of-mouth (WOM) marketing, so that if certain segments of consumers liked a product or service, they will share it with others virally via their mobiles and become part of the brand.

In this new type of social networking marketplace, friends represent a powerful new advertising model for interactive advertising via mobiles. Mining mobiles can be used to identify and target key "influencers" of products, services, and content. This type of engagement marketing is a fundamental switch from broadcasting marketing messages to consumers; to that of engaging them in conversations via their mobiles.

The social, user-generated content includes all types of mobile media, such as reviews, Facebook likes, Twitter tweets, and recommendations to other mobiles. To capture and measure these new engagement metrics, social media tools and platforms can be used, such as Bazaarvoice (http://www.bazaarvoice.com/; Figure 5.10) to track and quantify product ratings and reviews, as well as measure the content uploaded and shared via mobile connections on social networks. Bazaarvoice uses its MobileVoice service to deliver authentic product reviews directly to shoppers' mobiles in a store or mall.

The increased speed at which information is disseminated in a mobile world has made the patterns of the past less predictive of the future, and it is becoming

Figure 5.10   Bazaarvoice targets mobile shoppers.

increasingly important for mobile marketers to seek new research methods that measure mobile behaviors and communications. Social media research allows a brand or company to study large volumes of conversation at a lower investment and with faster turnarounds. Most important, social media research is not prompted or aided, but rather is based on the independent, autonomous, unfiltered, self-directed, and unaided conversations occurring in content created between consumers and their mobiles.

5.12 Mining Social Mobiles

Engagement mobile marketing uses influence metrics to measure a consumer's likelihood to encourage friends to consider, recommend, or purchase a brand via their mobile. In an increasingly mobile social Web world, texts, surveys, tweets, questionnaires, chats, apps, and instant messaging platforms provide new metrics for mining mobiles and the clustering of anonymous consumer categories based on the behavior of these devices. Brands would do well to listen to what consumers are saying about them, and the tracking of what brand advocates are saying via their product reviews or comments in social sites is vitally important for mobile marketers and enterprises that may want to enlist the services of peer reviewers, such as Epinions (Figure 5.11).

Engagement consumer profiles can be developed from passive, to semi-active participants, to the golden nugget of engagement marketing—the brand zealot—by the use of mobile analytics. Mining mobiles needs to focus on identifying these "brand zealots" in order to attract more with similar preferences, desires, demographics, and influences. Brands that have created strong emotional and personal associations tend to have higher attraction and loyalty in this era of social media, texts, emails, and other bursts of personal communications.

Mining mobiles can target evangelist consumers as co-marketers for their brand. Through the strategic use of engagement marketing, companies can invite consumers to become the co-producers of mobile marketing campaigns such as the creation of slogans, designs, widgets, videos, and other promotional viral advertisements.

Leveraging the communication and influence power of social networks for mobile marketers and developers is an important channel to be developed and utilized. The increasing prominence of social networks as the means to communicate and share information among their friends can be quantified by mobile marketers by enlisting ad agencies that focus on social media campaigns and metrics.

A new company, Factual™ (<http://www.factual.com/>; Figure 5.12) is doing just that; they are building databases from unstructured data whose primary focus is on basic information about businesses. Facebook is using Factual to populate its Places app to broadcast their locations to friends. However, more importantly, Factual is creating an ontology from this "unstructured data," as found in social media postings on the Internet, to automatically find and sort this data into

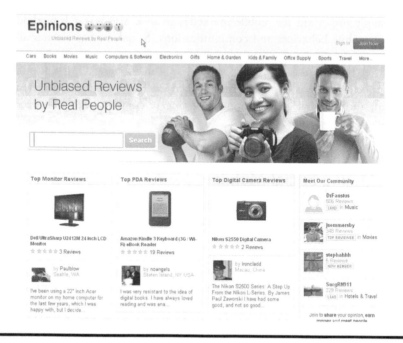

Figure 5.11   Epinions site.

Figure 5.12   Factual data site.

segmented categories; it then offers these segmented categories to companies that are eager to purchase it for mobile marketing apps.

Their Factual iPhone SDK helps developers and marketers quickly connect their iPhone apps to Factual datasets. Factual is an open data platform for app developers that leverages large-scale aggregation and community exchange. For example, the datasets have millions of U.S. and international local businesses and points of interest, as well as datasets on entertainment, retail, health, and education.

Factual provides a suite of simple data APIs and tools for developers to build mobile apps. In some cases, developers who create apps with Factual data may even get paid for crowd-sourced data from their users. Factual was originally Applied Semantics, which used natural language for the creation of anonymous consumer categories; they also developed AdSense, the contextual advertising platform that was acquired by Google.

5.13 An Artificial Intelligence (AI) Mobile

SohoStarCorporation (<http://sohostar.com/main.html>) has introduced a network and products that have AI capabilities, completely redefining the meaning of the "smart" phone. SohoStar's parent, Advanced Intelligent Networks Corporation, recently released David™, a highly extensible operating system for mobiles and all manner of digital communication appliances. David utilizes AI techniques to "learn" from a user's personal requests and creates a user profile applying factors such as age, interests, knowledge, etc. David keeps the data locally and serves social media, pictures, content, and email, providing personal information only to those whom the user designates.

David enables a live avatar that, with assistance from the hive of other David devices, provides VoIP Unified Messaging, Web and text communications, conferencing and multimedia streaming, as well as energy management and home controls. David provides anti-piracy content services and enables the collection of usage fees and honors encryption agreements for information rental.

With strict protections, the David built-ins enable running applications from anywhere on the planet on any David node—the true definition of cloud computing, software-as-a-service (SaaS). The software library shares information and protects the collective against cyber-attacks. David uses only 200 KB of firmware storage and can manage a device on its own or coexist with Linux. David bypasses the Java bloat of Android in order to be directly connected as the cloud.

SohoStar intends to provide global communication services through its network of National Franchises, the first being SohoStarRwanda, Ltd., with support from crowdfunding. David is an object-oriented, real-time operating system written in about 10,000 lines of C programming. Originally designed for very large telephony transaction processing systems, David contains a full LALR parser with multiple lexical analyzer tools, and a unique back-pass reduction system. David was designed to enable logic reduction algorithms to generate parser rules on-the-fly and modify the parser. This enables applications to be generated from dialog analysis.

David contains more than a hundred built-in data classes, including the X.208/X.209 ASN.1 classes. David combines schema definitions with object definitions to establish full database abilities without requiring full xSQL support. With the URI portal interface, David can utilize other system xSQL resources. David utilizes the AINC Neuron compiler that allows multiple programming language definitions, absorbing applications written in Java, Ruby, Smalltalk, C++, PHP, and other object-oriented languages.

David was designed for N+1 fault tolerance, allowing multiple systems to coordinate and load shared applications and services. The David architecture allows hot standby and resource recovery for distributed applications, especially for critical service systems such as 9-1-1 switches.

The David programming architecture includes the ability for Hot-Swapping live applications, including transfers of VoIP (Voice over Internet Protocol) processing in support of cell switching.

The David application installation interface provides full version, authoring, and resource management to search and resolve data definitions, classes, and applications, and ensure that linkage is specifically defined. In essence, if the installer cannot resolve all data definitions, library functions, and application interfaces, the application backs itself out.

Whether a successful load or not, David supports multiple resident versions of identical library functions. If the application ever works correctly, it will never degrade during revision updates. Because the system is intended for live network access, the system is always communicating with other David devices to share definitions, applications, interfaces, and services within the hive. Privacy and stability are paramount to the design of David. Multiple policies can be imposed for every facet of user, system, application, and global data, as well as multiple on-demand and escalation threat evaluation scenarios enabling throttling of actions and access as needed.

5.14 Business-to-Business (B2B) Marketing Analytics

With the incredible adoption of mobiles by businesses—to the tune of 174 million units shipped worldwide in 2011, according to Strategy Analytics (Strategicanalytics. com)—B2B marketers need to spend more time focused on their mobile marketing strategies, particularly as B2B mobile email open rates are among the lowest as compared to other industries via data mining mobile devices. Mobile-enabled emails cannot be smaller versions of standard communications and still be effective; marketers must understand the nuances of the mobile audience when tailoring campaigns accordingly. Here are five important points to consider when B2B marketing via email and analytics:

1. *Readability*: Emails should be easily readable on a mobile. However, because most emails contain links and offers, the developer and brand must ensure that their destination pages are mobile-optimized. Nothing is more frustrating to the reader than clicking through a link and ending up at a standard Web page form. The typical result is abandonment and brand erosion.
2. *Timing*: Companies and brands probably have tested for and identified the best time to send emails for optimal open/click-through rates; but, as mobile becomes ubiquitous, they might see those rates fall. Why? Because many business users do not open discretionary mail on their mobiles at the same time or in the same way that they do on their office desktops. Brands and companies might find that they have better success during off-hours when mobile users have the time to go through their inboxes. Test continually to determine optimal times for mobile or "blended" campaigns.
3. *Content*: Having unique, relevant content is more important than ever now that users are being bombarded by the same message in multiple platforms. By sending excessive, irrelevant, and redundant content, the mobile marketer can "repel" consumers. While generating new, fresh content can be challenging, it is important to keep an audience engaged across the mobile channel, which is immediate and cruel.
4. *Analysis*: Stating the obvious: know which customers are coming in on mobile devices. This intelligence allows the brand and enterprise to adjust how and where they should spend their precious marketing dollars.
5. *Testing*: If nothing else, the brand and company—it is critical in a mobile instantly enabled world to plan and test everything—must measure everything and use AI tools to model and predict how mobiles and their owners will react to content, cadence, timing, and continually optimize offers and content based on these results.

5.15 Mapping Mobiles

This type of undirected, unsupervised knowledge discovery, as previously mentioned, is best performed with neural network SOM programs. SOM software has been used for market basket analyses for years to discover who buys what, when, and where; however, such software has never been used in marketing to moving mobile devices. This is an important first task in data mining to mobile devices. Autonomous clustering is a data mining analysis—driven by the data, not the marketer or brand. The following are some of the most advanced clustering data mining companies in the field of AI:

■ *BayesiaLab:* This firm offers Bayesian classification algorithms to automatically cluster mobiles behavioral variables into distinct consumer groups. A

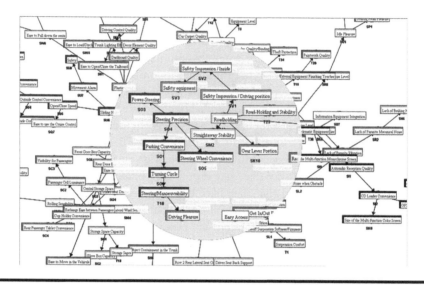

Figure 5.13 A Bayesian network can break down important features about mobiles.

Bayesian network is *a graphic probabilistic model* through which a user can acquire, capitalize on, and exploit knowledge. Particularly suited for *taking uncertainty into consideration*, a Bayesian network or belief network is a probabilistic graphical model that represents a set of random variables and their conditional dependencies (Figure 5.13).

For example, a Bayesian network could represent the probabilistic relationships between certain mobiles and consumer preferences. Given these preferences, a network can be used to compute the probabilities of the presence of Apple or Android mobiles and certain products or services being offered by a retailer or brand. A Bayesian network is used to analyze data in order to diagnose clusters and the resulting probability distribution of these effects. A Bayesian network is a graphical model that encodes probabilistic relationships among variables of location and interest and has several advantages for the discovery of hidden relationships for mining mobiles.

There are several advantages to the use of Bayesian network software. First, because the model encodes dependencies among all variables, it readily handles a situation where some data entries are missing, which is a common dilemma in data mining retail data. Second, a Bayesian network can be used to learn causal relationships and hence can be used to gain understanding about a problem domain and to predict the consequences of intervention—for example, what offers or deals to make to what mobiles. Finally, because the model has both causal and probabilistic semantics, it is an ideal representation for combining prior knowledge in large datasets for mining mobiles; in other words, the analyst or marketer knows best how to interpret the findings of a Bayesian network.

The software is extensive and no cause is overlooked, and all possibilities are considered. BayesiaLab is an ideal tool for analyzing the uses and attitudes of a group of customers, satisfaction questionnaire analysis, brand image analysis, segmentation, clusters and groupings of customers or products, and assessment of appetence scores in relation to new products and multiple mobiles. The BayesiaLab market simulator is a program through which a developer, brand, or marketer can compare the influence of a set of competing offers in relation to a defined population of mobiles, consumer behaviors, and groupings.

■ *Viscovery*: Viscovery (<http://www.viscovery.net/>) is the best neural network SOM clustering technology company in the world and is one of the most powerful unsupervised programs for knowledge discovery (Figure 5.14).

Mobile miners should not be aware of what this company and its founder Dr. Kramer offer; to do so will make them weaker to their competitors. SOM software such as that from Viscovery can be used for mining mobiles to target-group specification and campaign design based on historical mobile behaviors; the program can be used for customer segmentation and scoring,

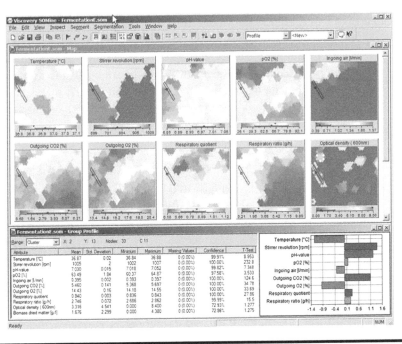

Figure 5.14 Viscovery SOMine can graphically show a mobile miner where the money is.

churn prevention, mobile key-performance indicator analysis, user mobile profiling, and the integration of empirical data. Viscovery provides a free evaluation of their Austrian software.

■ *PolyAnalyst*: PolyAnalyst (<http://www.megaputer.com/site/polyanalyst.php>) is a Russian-based software from Megaputer, which offers its clustering analysis based on a Localization of Anomalies (LA) algorithm. They offer more data mining algorithms in their software than practically anyone else in the industry, including SPSS (now IBM) and SAS, the behemoth of data analytics. Megaputer has a dynamic software product called X-SellAnalyst™ that empowers mobile sites to recommend new products in real-time to mobile visitors. X-SellAnalyst offers products or services that fit the needs of individual visitors to build customer loyalty.

X-SellAnalyst trains itself on available transactional data and develops and stores a "recommendation filter" that can be used for quickly calculating the most attractive cross-sell opportunities. Once trained, X-SellAnalyst can serve as an intelligent adviser, recommending only the products that have the best chance of being purchased. X-SellAnalyst is extremely scalable; it can train quickly on data involving dozens of thousands of products and millions of customers and transactions. X-SellAnalyst can be easily incorporated into any mobile site; it is packaged as a *Component* Object Model (COM) module. One of the features of X-SellAnalyst is its scalability; the calculation time for 100,000 products has a recommendation time of milliseconds.

■ *perSimplex*: perSimplex (<http://persimplex.biz/>) is clustering software that is based on fuzzy (continuous) logic; the program can be used to find hidden connections in mobile data. It is software based on natural shape analysis: fuzzy logic. perSimplex is able to process interaction and behavioral data via visualization designed to influence a mobile marketer's decisions; it does this by generating curve shapes to clarify hidden relations within mobile activities (Figure 5.15).

perSimplex selects the relevant pieces of information from huge volumes of data sources and creates clustering analyses based on the natural shape of data curves using fuzzy logic technology. Fuzzy logic is a form of reasoning that approximates behaviors rather than fixed logic or exact shapes; it replicates the way humans reason by approximation rather than by crisp black or white inference; and it is continuous logic.

perSimplex is fast and simple with the ability to identify natural data clusters and data abnormalities. The fuzzy logic software can identify adequate clusters and has the ability to distinguish the level of similarity and to identify the abnormalities of input data via its proprietary nonhierarchical cluster algorithm for linear computing complexity. In other words, by using this kind of program, segmentation of app users or mobile site visitors can be approximated in a very granular and highly accurate manner by developers, marketers, or brands.

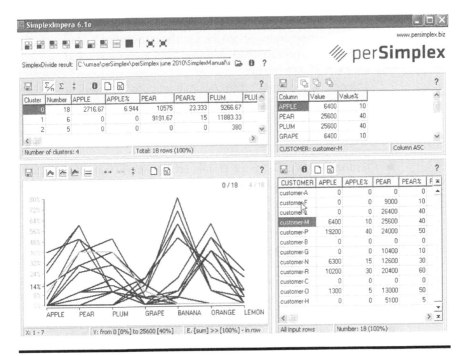

Figure 5.15   perSimplex mapping clusters of mobiles.

Case Studies: Mobile Analytics

Shopkick™: Shopkick (<http://www.shopkick.com/>) is the first mobile app that gives rewards and offers for simply walking into a shopping area. Further functionality allows customers to get additional rewards for scanning products and trying on clothes. It is a way to incentivize activities that retailers know increase on-site purchases. What is most significant about the app's rollout is the all-star roster of firms that have already endorsed it. Macy's, Best Buy, American Eagle Outfitters, and Sports Authority are rolling out Shopkick at their stores, as is Simon Property Group, the largest regional mall owner in the United States.

Shopkick is currently active at about twenty Simon malls in New York, Los Angeles, San Francisco, and Chicago. Simon plans to roll out the application at more than 100 of its properties. The app, which has been about a year in the making, is meant to encourage more store visits. Stores that want to use Shopkick install a location device on-site that emits a silent audio signal when a customer with the app on his/her mobile walks inside the store. At Simon properties, the devices have been installed at the entrances to the malls, as well as in food courts and common areas. Shopkick's audio transmitter is several degrees more precise than standard GPS technology and costs less than $100 to install.

Usablenet: Usablenet (<http://usablenet.com/>) offers a leading technological platform for transforming and optimizing mobile site content and apps; the following are some of their clients:

- *Fairmont Hotels*: Fairmont Hotels (<http://www.fairmont.com/>) customers are, by definition, travelers; being able to reach them on mobiles is critical to building closer relationships with them and bridging the gap between inquiry and reservation. Fairmont Hotels contracted with Usablenet to create an app that would enable frequent travelers and preferred customers to view photo slideshows of Fairmont properties, search for hotel packages by theme, book or change reservations, and learn of nearby attractions and activities.
- *Garnet Hill*: Garnet Hill (<http://www.garnethill.com/>) began as an importer of English flannel sheets and has grown into a distinguished brand and multichannel marketer. Before launching its mobile site, 80 percent of users surveyed were unable to complete a product purchase and 45 percent of them abandoned the transaction due to the site's inability to load. Garnet Hill rolled out its mobile commerce site, built by Usablenet and within two months, the company saw a 300 percent increase in mobile sales.
- *PacSun*: PacSun (<http://www.pacsun.com/>), a leading supplier of clothing and accessories for fun in the sun, is pressing ahead aggressively on a number of fronts to make it the first choice of hip, active men and women. The company has been selling its products through an optimized mobile commerce site and engaged with Usablenet to create a full-featured app that enables customers to buy any product within the app.

Placecast: Placecast (<http://placecast.net/>) works with many brands via their ShopAlerts service, which is an opt-in, location-based mobile marketing solution that delivers highly relevant SMS messages customized for each customer based on their location, time, user preferences, and CRM data. Their clients include the following:

- *The North Face:* Summit Signals, The North Face location-based messaging program, enables fans of the brand to stay connected about the latest gear, local events, sponsored sports events, and tips for outdoor activities.
- *White House Black Market*: Fashion Alerts, the location-based mobile program from White House Black Market, offers the brand's customers exclusive offers and promotions as well as the insider scoop on new product arrivals and VIP store events.
- *Sonic*: SONIC Drive-In customers participated in the restaurant's Sonic Signals programs and received special promotions and offers when they enter geo-fences created areas around SONIC restaurants.
- *O2*: Placecast's collaboration with O2, the second-largest mobile operator in the United Kingdom, enables ShopAlerts to be utilized directly by the carrier for more than twenty brands and advertisers and more than one million opted-in subscribers.

MobilePosse: MobilePosse recently worked with Ford to deploy idle-screen ads via mobiles to help drive interest in its Taurus model. The idle-screen ad racked up an enviable 20 percent average click-through rate. Working with Mobile Posse Inc., a provider of mobile marketing services, Ford increased purchase consideration and referred consumers to the Taurus mobile site, where they could view product videos, find a dealer, and receive more information.

Medialets: Medialets worked with several clients to develop mobile-rich media marketing campaigns; their clients include the following:

- *HBO:* This innovative app ad utilizes a dramatic approach to promote the HBO *True Blood* series, employing tactile features that leave bloody fingerprints as the user interacts with the screen, followed by a transparent overlay that shows blood dripping down and a link to their trailer video. The creative generated extraordinary buzz among *True Blood* fans and the advertising community, and was broadly recognized for excellent mobile-rich media. For mobiles, this is a critical marketing strategy, involving social media, entertainment, and a spin on a hot TV program for young consumers.
- *JPMorgan Chase:* Medialets created and built a rich media mobile ad featuring a Chase Sapphire card that,

when tilted, "spilled" out rewards and links to their mobile site where users could learn more about rewards and the card.

■ *BayesiaLab:* The mobile databases relative to the customers can be used to elaborate profiles using data mining methods. These profiles can then bring objective information to the marketing department; they can also be used to reduce the cost of the campaigns by selecting only the prospects that have a high probability to reply positively. BayesiaLab was used for the profiling of customers with respect to a bank mobile product that had various modalities. The database used contained various variables describing the customer, such as age, socio-professional, gender, etc.; it also had bank account variables such as facilities, rate of consumption, etc., and the modality of its bank products.

Bayesian networks represent all the probabilistic relations that can be held in a data set; for example, their Markov Blanket Learning algorithm allows focusing a search only toward the variables that really characterize the target mobile. The Bayesian network could be used to predict a bank product for new customers. As BayesiaLab returns the probability associated to that prediction, it is possible to use their probability to reduce the cost of marketing campaigns by using the most printable targeted mobiles

Case Studies: Banking Mobile

Banks are increasingly seeking a core system that utilizes data analytics to provide a more complete view of their customers, according to a new report from the Boston-based analyst firm Celent. The report, titled "Core Banking Solutions for Midsize Banks: A Global Perspective" (<http://www.celent.com/reports/core-banking-solutions-midsize-banks-global-perspective-0>), examines the core solutions used by a select number of banks with between $1 billion and $20 billion in assets, the future of which is continuously mobile.

In addition to increased analytics capabilities, Celent found that banks are also seeking a multichannel, integrated core banking solution. Multichannel options such as ATM, mobile, Internet, and IVR (Interactive Voice Response) are crucial in attracting and serving customers, the report reads. Core banking solutions are also expected to facilitate product development and provide flexible customization capabilities.

The report also noted that there is a "major preference" for in-house implementation. Especially in developing countries, banks prefer in-house implementations to hosted implementations. In many cases, there is no hosted option in a given country, Celent noted; also noticed was a decrease in demand for mobile core banking replacements in the current economic climate. Banks switching core banking solutions are relatively higher in South Asia, Middle East Asia, and Africa, however, where the telecommunication infrastructure does not exist—but mobile can waltz in. The major trend is that mobiles will be the future for how a vast majority of the world will check their checking account balances and look at their mortgage and saving accounts.

CART®

Fleet Financial Group, a Boston-based financial services company with assets of more than $97 billion, is currently redesigning its customer service infrastructure, including a $38 million investment in a data warehouse and marketing automation software. To profit from this repository of valuable information on more than 15 million customers, Fleet analysts used Salford Systems CART® (<http://www.salford-systems.com/en/products/randomforests>) to learn about their customers and to better target product promotions, such as home equity lines of credit increasingly going mobile.

To do that, Fleet needed to learn about customers' financial characteristics and buying habits so as to target the mailing list for the company's third-quarter home equity product promotion. The first step in the modeling process was to gather historical data on which to create the data mining model. The team selected a sample of approximately 20,000 customers for which Fleet had a record of responses; included were 100 percent of past profitable respondents, as well as 2 percent of past nonrespondents.

The dataset was then transferred into CART to display the interaction of the data. When the data was fed into CART, the software automatically generated a decision tree whose branches, or nodes, showed the hierarchy of binary data splits and displayed the dataset's myriad variables and their interactions. This hierarchy distilled nearly one hundred predictor variables into a more compact number of approximately

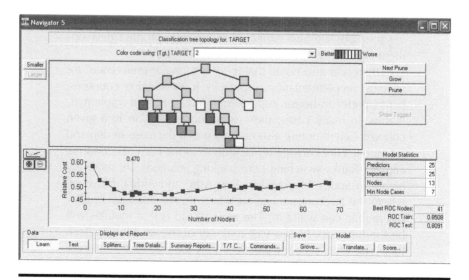

Figure 5.16　CART in action.

twenty-five. In addition, the CART nodes provided probability ratios that can be used to understand why one segment would be more responsive than another (Figure 5.16).

The CART model illustrated certain characteristics of "best" respondents by predicting the expected balance they would carry on the credit line, as well as how much they might transfer from another line. In addition, the CART results painted a portrait of the principal characteristics of the least responsive customers. These prospects would either not likely respond to a Fleet product offer because they do not have a need for a large line of credit—or of equal concern, they would respond, but their subsequent credit line usage would not be profitable for the bank. Fleet continues to use CART to gain a deeper understanding of its customers so that the information can be applied to classification and segmentation applications.

STREAMBASE

BlueCrest Capital Management is a leading European hedge fund based in London, with over $15 billion in assets-under-management. They set up a team to develop a state-of-the-art market data management system. BlueCrest trades twenty-four hours a day, six days a week, across multiple markets using a wide range of data feeds. BlueCrest needed to rapidly plug the data into real-time models while optimizing management of the necessary data feed licenses. They devised a solution

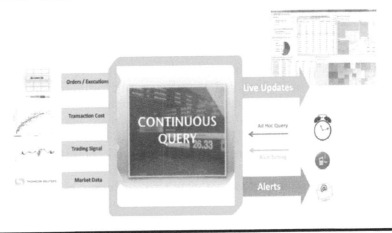

Figure 5.17   Streaming analytics of mobile data.

that combines the rapid time-to-market event processing using StreamBase (<http://www.streambase.com/>; Figure 5.17).

- *Visipoint*: Visipoint is SOM software for autonomous clustering and visualization. Aside from organizing key concepts, this clustering software can be used to self-organize mobile behaviors. Mobiles can also be clustered by text analytics software into key segments along service and product lines. For example, visitors to a mobile site can be grouped into those who are interested in sports, financials, music, or other major topical categories.

5.16 Text Analytics

Text analytics, sometimes alternately referred to as text data mining, refers to the process of deriving high-quality information from unstructured content, such as tweets, emails, and online comments. High-quality information is typically derived through the devising of patterns and trends through means such as statistical pattern learning. Text analytics usually involves the process of structuring the input text—usually parsing, along with the addition of some derived linguistic features and the removal of others, and subsequent insertion into a database—deriving patterns within a structured dataset, and finally evaluating and interpreting the output.

Text analytics usually refers to some combination of relevance, novelty, and interestingness from mobile behaviors and preferences. Typical text mining tasks include text categorization, text clustering, concept/entity extraction, production of granular taxonomies, sentiment analysis, document summarization, and entity relation modeling (i.e., learning relations between named entities). Text analysis involves information retrieval, lexical analysis to study word frequency distributions, pattern recognition, tagging/annotation, information extraction, and data mining techniques including link and association analysis, visualization, and predictive analytics (Figure 5.18). The overarching goal is, essentially, to turn text into data for analysis via application of natural language processing (NLP) and analytical methods.

A typical application is to scan a set of texts, comments, blogs, or documents written in natural language and either model the document set for predictive classification purposes, or populate a database or search index with the information extracted. The following are some of the major text analytical software products; they are relatively easy to use and can yield important results for mining text from

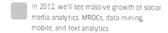
In 2012, we'll see massive growth of social media analytics, MROCs, data mining, mobile, and text analytics

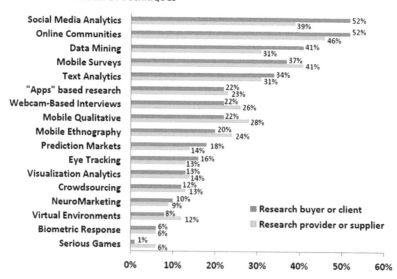

Figure 5.18   Text mining findings from mobiles.

mobiles by performing data analyses of unstructured content for consumer sentiment categorization, emotional brand, social and mobile intelligence gathering.

■ *ActivePoint®* (<http://www.activepoint.com/>): Text mining software for interactive search and product suggestion, their patented TX5 system is a contextual dialogue-orientated natural language system that is combined with engines that utilize rule-based reasoning and algorithms. The TX5 systems are able to automate mobile sites. The software's uniqueness is that it has the ability to not only find products, but also to educate, convince, suggest relevant products, answer consumer questions, compare products, handle objections, and offer alternatives. It can encourage mobiles to make purchases, mimicking a live sales assistant.

■ *Aiaioo Labs* (<http://twitter.com/aiaioolabs>): Text categorization and sentiment analysis software. Mobile reviews and messages of products or services often express positive or negative sentiments that are valuable to collect and summarize. This task requires the classification of an utterance about products or services into positive, neutral, or negative utterance. This task is known as *sentiment classification;* this is what their software can do for mobiles.

■ *Attensity* (<http://www.attensity.com/home/>): Can extract "who," "what," "where," "when," and "why" and their relations to each other directly from text. Attensity provides a suite of integrated applications that allows mining mobiles to extract, analyze, quantify, and act on customer conversations from unstructured content. The company provides hosted, integrated, and stand-alone text analytics solutions to extract facts, relationships, and consumer sentiment from unstructured data emulated from mobiles (Figure 5.19).

■ *Clarabridge®* (<http://www.clarabridge.com/>): Software can perform sentimental and concept extraction; it can transform text-based content into quantitative and easily consumed reports and analysis from apps or mobile sites. Clarabridge extracts linguistic content, categorizes, and assigns sentiment scores to distinguish the "who," "what," "how," and "why" of any customer experience; it makes data accessible from a variety of interfaces for marketers and brands.

■ *ClearForest* (<http://www.clearforest.com/>): Derives meaning from mobile sites, blogs, tweets, emails, texts, etc. Their OneCalais algorithm uses Natural Language Processing (NLP), text analytics, and data mining technologies to derive meaning from unstructured information. The software categorizes each piece of content using both data streams with hidden codes and social tags. The FBI uses the software to monitor mobiles for counterintelligence analyses.

■ *Crossminder* (<http://www.crossminder.com/>): A natural language processing tool for analyzing mobile sites, texts, and emails, this software helps automate the processes where languages and the meanings of unstructured content play a central role. The software can identify market trends, interpret

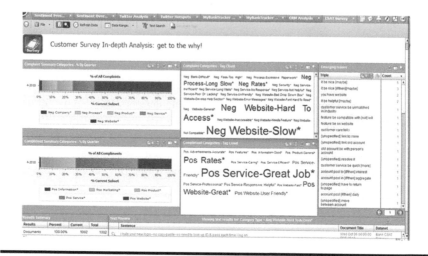

Figure 5.19 An Attensity customer analysis text mining report.

and predict mobile behaviors, and develop lexical terminological categories, such as those that "like" and those that do not.

- *dtSearch®* (<http://www.dtsearch.com/>): Fast indexing, searching, and retrieving software that can organize millions of text files. The dtSearch product line can instantly search terabytes of text across mobile networks and mobile sites, and organize them into insightful categories for marketers.

- *Lexalytics* (<http://www.lexalytics.com/>): Converts unstructured text for social network monitoring. The software can automatically extract companies and brands and understand the sentiment (tone) being directed toward each; it does this across any type of text. Lexalytics uses thematic analysis to understand the concepts present in conversations; it uses these themes to understand the words that people are naturally using to convey their view of brands. Mechanically processing sentiment across thousands of concepts and entities leads to a much greater understanding of the semantic map of potential bias, the software can be integrated into mobile apps and sites to obtain metrics on brand loyalty.

- *Leximancer* (<https://www.leximancer.com/>): Their software is presented as a "Concept Map" depicting the salient concepts and their relationships, which can be readily interrogated at either the theme or concept. Sentiment is becoming a paramount concern for mobile marketers, especially in today's interconnected environment. It can answer what mobiles are saying about brands and why.

- *Nstein* (<http://www.nstein.com/>): A semantic site for mobile site content management; their Text Mining Engine (TME) helps extract the value from content, and maximize it, while reducing content-related costs. With its patented algorithm, Nstein creates a "semantic fingerprint" for easy identification, association, and to identify nuance and meaning in content for brands.

- *Recommind*® (<http://www.recommind.com/>): Using a probabilistic latent semantic algorithm, their tool automates categorization. Recommind is powered by its context optimized relevancy engine (CORE) platform, a fully integrated set of technologies that delivers the most accurate information, irrespective of language, information type, or volume for concept-search and categorization. This type of software can be used by brands with international mobile sites and apps.

There is also machine-learning software for consumer device segmentation, classification, and prediction. These types of programs can generate graphical decision trees, as well as predictive rules about mobile behaviors. The following section contains some of the most powerful and effective behavioral, deductive, predictive, modeling, and extremely precise classification software.

5.17 Behavioral Analytics

The fledgling mobile transactions market, which some experts expect to reach $1 trillion globally by 2015, is about to get a hyper-boost from Facebook, Twitter, Square, and other social media players that consider mobile sales the new end-game. Despite discrepancies among forecasters, mobile transactions are clearly morphing into a critical revenue stream for brands, retailers, and product and service providers. The 15 percent leap in online holiday spending to more than $35 billion, and the overall retail e-commerce growth of 13 percent to $161.5 billion in 2011, according to comScore (comScore.com), were driven by consumers' accelerated use of mobile price comparison and payment apps.

This activity has been fueled by Square, which progressively facilitates mobile consumer transactions by acting as a medium between merchants and payment networks—credit-card companies and banks—across all mobile devices. In addition to securing consumers' personal information and providing effective user discount incentives, Square creatively uses hyper-local and social media app components, all of which are driving a flood of competing mobile payment solutions.

The mobile transaction opportunity is huge, according to Wedbush Securities (Wedbush.com), which estimates that the transaction volume across Square's four payment networks (Visa, MasterCard, AmericanExpress, and Discover) was more than $6 trillion in 2010.

Little wonder that Facebook and Twitter are positioned for a bigger piece of the action by way of an aggressive mobile ad grab, which the social networks consider a means to a transactional end.

Facebook is leveraging off of its "like" and "own" buttons and new Timelines by inserting featured stories—or relevant marketing-inspired posts—into mobile feeds that are squarely aimed at its 425 million active mobile users. Facebook recently announced a partnership with UK mobile billing and analytics provider Bango (<http://bango.com/>), considered a step toward monetizing and expanding a browser-based mobile platform beyond its Facebook credits program.

While the domestic mobile ad market is an estimated $2 billion, it is a sure stepping stone to the larger mobile transaction market, which can be incrementally shared by all players. Facebook is playing off its competitive share of online display ad revenues in the United States and the United Kingdom and reliance on small businesses. It must make a major play for mobile transactions to justify its proposed $100 billion public valuation.

Competition for the data mining of mobile device transaction market is coming from many corners, including Twitter, which announced an expansion of its self-serve advertising program with American Express as a partner. The goal is to attract the small-to-medium local businesses where mobile transactions and marketing can be a big win-win for all concerned. The program opens up to anyone with a credit card later this year.

Location-based interactivity will transform advertising and marketing into pre-cursors to secured, single-click sales.

Social networks are serious contenders for the data mining of mobile devices and strategically advertising and commerce dollars; they account for one in every five minutes online globally and reaching 82 percent of the world's Internet users, or 1.2 billion users, according to comScore (comScore.com). With transactions being the ultimate objective for all marketers, it is likely that by decade's end, Facebook and Twitter will be vying for more of television's $9 billion up-front ad pie along-side Google and Hulu, which are aggressively positioning themselves this spring with the Big Four TV networks.

The direct line to consumers adeptly secured and mined by Facebook will give it an edge as marketers increasingly realize transactional revenues as a measured ROI (return on investment). A recent Yankee Group survey revealed that consum-ers value banking and transaction apps second only to social networking apps as the most important on their mobiles, a simple but impressive indication of just how comfortable they have become with using these powerful pocket computers (mobile devices) for secured personal transactions.

It is a sure sign of the next phase of connected growth. What we have seen so far has been all about consumers learning to navigate their lives with mobile devices. What we are about to see unfold is how consumers use connectivity to monetize their lives. The race is on for a piece of that new economy. Following are some of the software tools for data mining mobile devices:

- *AC2*: AC2 is a graphical tool for building decision trees. It allows the creation of segmentation models; it generates classification reports and cross-tables. The software generates utilitarian and dynamic measures of dependencies

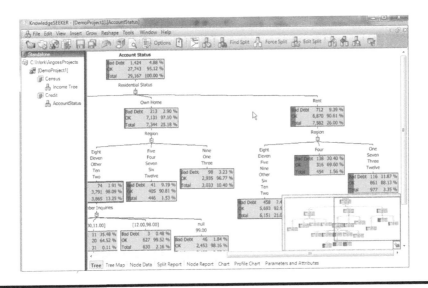

Figure 5.20   A decision tree for targeting mobiles.

across data variables. It regroups similar individuals into classes of certain types of mobile behaviors; it compares numerous data sets and makes mobile marketing forecasts (Figure 5.20).

▪ *Angoss KnowledgeSEEKER*: Multiple-branch tree and rule generator software for segmentation. This is one of the most powerful, oldest, and robust classification programs in the marketplace. For the mobile marketer or developer, it is a very effective and precise mobile mining tool; it uses four algorithms based on variants of CHAID and CART algorithms. The software has an advanced algorithm control and user preferences to fine-tune tree growth and can generate code in structured English, generic, SAS, SPSS, SQL, Java, PMML, and XML, code that can be incorporated into apps and mobile sites to improve relevancy, revenue, loyalty, and brand.

▪ *C5.0*: Constructs classifiers in the form of decision trees and predictive rules. This program introduced the "information gain" algorithm and has been designed to analyze numeric, time, date, or nominal fields; it was created by Dr. Ross Quinlan, the author of the ID3, C4.5, FOIL, and C5.0 algorithms.

▪ *CART*: Creates compact decision trees for predictive modeling of mobile behaviors. It is robust, easy-to-use decision tree software that automatically sifts through large, complex databases, searching for and isolating significant patterns and relationships for predictive models for apps and mobile sites, such as finding the best prospects and customers via the behaviors of their mobiles.

▪ *DTREG*: Can generate classification and regression decision trees as an ensemble of predictions combined to make a "decision forest" (Figure 5.21). A decision forest grows a number of independent trees in parallel; their models

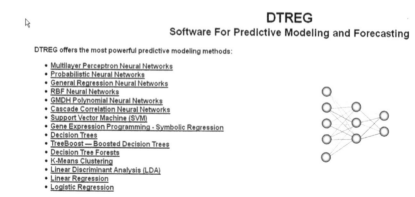

DTREG

Software For Predictive Modeling and Forecasting

DTREG offers the most powerful predictive modeling methods:

- Multilayer Perceptron Neural Networks
- Probabilistic Neural Networks
- General Regression Neural Networks
- RBF Neural Networks
- GMDH Polynomial Neural Networks
- Cascade Correlation Neural Networks
- Support Vector Machine (SVM)
- Gene Expression Programming - Symbolic Regression
- Decision Trees
- TreeBoost — Boosted Decision Trees
- Decision Tree Forests
- K-Means Clustering
- Linear Discriminant Analysis (LDA)
- Linear Regression
- Logistic Regression

Figure 5.21 DTREG multiple algorithms.

have a high degree of accuracy that cannot be obtained using a single-tree model, and forests can construct the most accurate models yet invented in the fields of machine learning and AI.

■ *PolyAnalyst*: Includes an information gain decision tree and other algorithms, from Megaputer. It supports data preprocessing and modeling via prediction, classification, clustering, affinity grouping, link analysis, multidimensional analysis, and interactive graphical reporting of mobile behaviors.

5.18 Streaming Analytics

The data mining of mobile devices may require the use of both deductive and inductive "streaming analytical" software that is *event driven* to link, monitor, and analyze mobile behaviors. These new streaming analytical software products react to mobile consumer events in real-time. There are two main types of streaming analytical products:

1. There are deductive streaming programs that operate based on user-defined business rules and are used to monitor multiple streams of data, reacting to consumer events as they take place.
2. There are also inductive streaming software products that use predictive rules derived from the data itself via clustering, text, and classification algorithms. These inductive streaming products build their rules from global models involving the segmentation and analysis from multiple and distributed mobile data clouds and networks.

These deductive and inductive software products can work with different data formats, from different locations, to make real-time predictions using multiple models from massive digital data streams.

■ *Progress*®: Enables adjustments to changing mobile interactions as they occur. Mobile events flowing in and out of a mobile site or an enterprise can be subtle, such as an app entering a Wi-Fi hot spot or being triangulated when a mobile enters a geo-fenced location in a store or mall. The ability to respond in real-time to these consumer mobile events can provide a competitive advantage to the marketer, retailer, or brand.

The speed of advertising and the ever-changing marketing conditions means that traditional methods of handling behavioral mobile data do not go far enough. The past-tense nature of existing marketing solutions to human consumers often results in too little relevant targeting intelligence delivered to their mobiles. The Progress® Apama® Event Processing Platform can provide the ability to monitor the activity of mobiles via its dashboard providing up-to-the-second visualization of key performance indicators (KPIs). This streaming analytical software has the ability to see operational activity in real-time for generating offers of products and services to moving mobiles. The processing platform monitors rapidly moving event streams, detects and analyzes important patterns, and acts in sub-millisecond time, whether location or interest based. With this type of streaming software, the marketing of mobiles events can be analyzed, providing new levels of decision making that dramatically improve the precision of data mining mobile devices.

■ *Splunk* is streaming analytics software that collects, indexes, and harnesses all of the fast-moving mobile data generated by apps, servers, and devices—physical, virtual, and in the cloud. It troubleshoots application problems and investigates security incidents in minutes instead of hours or days, avoids service degradation or outages, delivers compliance at lower costs, and gains new business insights.

Strategic Vision

Mobile is hot: the VCs have dipped into their wallets and doled out big bucks to mobile-focused companies. Even more than just mobile, there is a veritable gold rush into the mobile commerce market from mobile payments to mobile couponing to location and interest-based discounts to the MoLoSo model—Mobile-Local-Social—integrated technologies. For CEOs, brands, and other enterprises, the big question is what they really need to know about data mining mobile devices right now. Here is a listing of essential things to consider:

First and foremost, brands and developers should know that mobile commerce is already here. According to eMarketer (eMarketer.com), over 50 percent of Americans own mobiles, with the biggest market share in the hands of Apple iOS and Google Android platforms, and experts expect that by 2015, more than 8 percent of the overall e-commerce spending will occur on mobiles.

While those are staggering figures for such a "youthful" market, let's also remember that mobile commerce would still only be half a percent of total retail sales. Mobile commerce and innovation are still in their infancy. The concept of the "app" is less than five years old. While the technology and the market are going gangbusters, it is still an early market with much left to be sorted out, including the data mining of mobile devices for brands and marketers.

Consumers are carrying these powerful mobile devices in their pockets and using them to make buying decisions; already 79 percent of consumers are using their mobiles to assist them while shopping. That number will grow exponentially in the coming years. For example, Starbucks is the leading company in terms of mobile payment, transactions, and dollars in the world. The Starbucks iPhone and Android apps let their customers check their account balance, reload the balance on the app with any major credit card, and view past transactions. Customers can also place their order from their mobiles, receive a barcode onto their device and scan the barcode at the register to pay, and more and more retailers will follow this retail model.

This seemingly simple transaction of "buying coffee from your phone" has introduced millions to mobile commerce already. And buying coffee is not the only way people are trying out mobile commerce. Shoppers are using their mobiles while browsing the aisles to assist them with their shopping. As amazing as this might sound, more than 31 percent of male and 23 percent of female mobile users said they had spent more than $500 on mobile commerce in the past twelve months.

In a survey by Zaarly (Zaarly.com) of over 2,000 online shoppers, less than 25 percent of all online shoppers have ever purchased anything from their mobile phone. Expect those numbers to grow substantially as companies like Starbucks introduce more and more consumers to the concept of paying for real-life goods or services on their mobiles and consumers experience more shopping directly from their devices.

Mobile commerce will remain a novelty for many new mobile shoppers, but expect to see more consumers make the move from novelty mobile shoppers into regular mobile purchasers. There are payments giants such as Visa and Amex, and tech giants such as Google, Paypal, and Intuit, and start-ups such as Square and Dwolla. Recently, one of the hottest mobile apps has been Path, which was beset with a privacy scandal when pundits uncovered that Path was accessing the mobile address book and transferring that information from the phone to its servers. To its credit, Path quickly addressed the issue and updated its app within hours of the issue. But the issues faced by Path are becoming the norm, having arisen for many a company—from Facebook to Google. But be prepared to uncover a new world unlike existing online advertising platforms: lots of players, not a lot of transparency and analytics, and unclear ROI.

Lots of companies are hoping to build their own empires around mobile advertising. It is a different ballgame on mobile with the mixture of apps, mobile Web, and traditional Web, coupled with varied real estate on mobile devices. But look for consolidation and more offerings from "consultants" or "aggregators" willing to help build, track, and optimize mobile advertising across the various mobile channels.

"Mobile commerce" is here and 2013 will be an important year as brands and marketers determine what the heck that even means. Does the mobile represent a new wallet, or a new shopping companion, play games, check email, and meet up with friends? Smart CEOs and brands need to keep their business plugged in on mobile and make smart bets about the exciting future ahead.

───────────────────────────────

Splunk is streaming software that collects, indexes, and harnesses all the fast-moving machine data generated by apps, servers, and mobiles—physical, virtual, and in the cloud. Splunk is a leading provider of operational intelligence software used to monitor, report, and analyze real-time machine data as well as terabytes of historical data located on-premise or in the cloud. Almost half of the Fortune 100 and more than 2,300 enterprises, service providers, and government organizations in seventy countries use Splunk to improve service levels, reduce IT operations costs, mitigate security risks, and drive new levels of operational visibility. Splunk aims to make machine data accessible across an organization and identifies data patterns, provides metrics, diagnoses problems, and provides intelligence for business operations. Splunk is a horizontal technology used for application management, security, and compliance, as well as business and mobile analytics. Splunk has over 3,700 licensed customers.

The *Sybase Aleri Streaming Platform* is streaming analytic software that supports event-driven analytics from user-defined rules. Delayed response to changing conditions can mean missed opportunities from streams of incoming mobile behaviors. The Sybase Aleri Streaming Platform enables rapid application development and deployment of real-time marketing applications that derive insight from streaming event mobile data, empowering instant responses to changing conditions. The Aleri Streaming Platform consists of a high-performance Complex Event Processing engine and the Sybase Aleri Studio, for rapid application development and a range of integration tools including adapters and APIs.

Streambase is another complex event processing (CEP) platform enabling the rapid development, deployment, and modification of applications that analyze and act on large amounts of real-time streaming mobile data. CEP is a technology for low-latency filtering, correlating, aggregating, and computing on real-world event data. The software can be used for rapidly building systems that analyze and act on real-time streaming mobile data for building microsecond marketing systems.

Palantir is yet another streaming firm with a high-powered analysis platform that can scan multiple databases simultaneously, a tool that government officials and corporations use to tackle complex problems, especially for data mining mobile devices across disparate databases, networks, and clouds. Palantir is not in the social game. It neither dispenses daily deals, nor does it accept mobile payments—and it certainly does not "tweet."

It is an obtuse, difficult-to-explain product that is mainly used in Washington for counterterrorism; the government makes up 70 percent of its business and the rest is dominated by private financial institutions. That may sound painfully boring, but Palantir's user-friendly analysis program is becoming a major player in the war against cyber-espionage, stimulus spending accountability, fraud detection, health care, and even natural disasters such as the recent earthquake in Haiti. But its true value will be in streaming data mining mobile devices.

Wavii automatically generates status updates for celebrities, companies, politicians, and other topics. Wavii accomplishes this with their proprietary NLP technology that reads articles and blogs across the Web, de-duplicates each story, and figures out what is happening with these topics. This is streaming analytics *text* software (Figure 5.22). For example, a political victory, breakup, acquisition, or new product releases.

InferX is inductive streaming software that develops rules constructed from multiple distributed data sources. This software can be used for mining mobiles from multiple, disparate, and remote databases, servers, and networks requiring the use of a software agent to access, analyze, and perform predictive analyses in real-time. This software works on multiple data formats and involves a number of independent Modeling Agents (MAs) operating within a network-based environment. Each MA operates on its own local database and is responsible for analyzing the data contained by that dataset.

Distributed data analytics is achieved through a synchronized collaboration of MAs facilitated by a Collaboration Server (CS) environment. This process results in

iPhone Screenshots

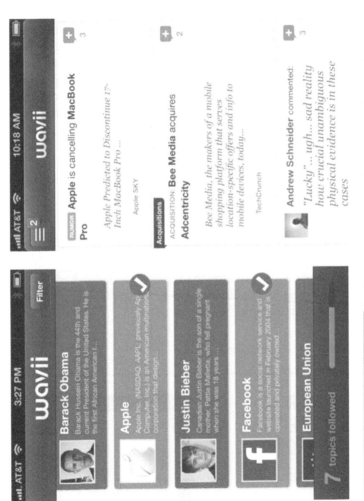

Figure 5.22   Streaming social media mobile data by Wavii.

a set of global rules generated by the partial contribution of each MA and assembled through a CS communication mechanism. Collective decision making is similarly achieved through the collaboration of Decision Agents (DAs), operating independently on their own data.

Metamarkets is another event-driven firm. Companies are generating massive amounts of event-based data, such as impressions, tweets, check-ins, purchases, actions, transactions, payments, and other meaningful events. Analyzing and understanding these data streams can increase revenue, improve user engagement, and increase operational awareness.

However, very few are able to take advantage of this opportunity. Current tools and processes are inadequate, and most companies lack the highly skilled data scientists needed to turn data into useful information. The Metamarkets mission is to democratize data science by delivering powerful analytics that are easy and intuitive for everyone. The company's leading-edge analytics platform delivers real-time insights. Customers are able to get started quickly and scale easily using the Metamarkets cloud-based solution.

The Metamarkets analytics solution (Figure 5.23) is built on a big data stack for processing, querying, and visualizing high-volume, high-frequency event streams. Its components include

- Data Pipes: Their customized Hadoop pipeline for parallel data processing.
- Druid: Their distributed, in-memory data engine that can slice, dice, and drill through data 1,000 times faster than conventional, disk-backed databases.
- DVL: Their dynamic visualization LEGOs, a JavaScript framework for interactive data visualizations.

Figure 5.23 **Metamarkets streaming analytics.**

5.19 Mobile Site Analytics

Increasingly, reporting and analyzing the behaviors of devices at mobile sites is crucial to brands and marketers. Web analytics software can help enterprises measure the traffic and behaviors of mobiles at their sites. Mobile Web analytics provide information about the number of visitors and page views to a mobile site as well as the path taken by them. They help gauge traffic and popularity trends, which is useful for data mining mobile devices. The following are some of the most popular mobile Web analytical software products and services:

- *AlterWind:* Log analyzer with search engine optimization and website promotion. This software has its own unique analysis code of the URL for each search engine, taking into consideration the specialties of each search engine. This will help the mobile marketer gather valid information, more clearly understand the interests of mobile site visitors, and receive additional data for website search engine optimization. Their professional version has a database with more than 430 search engines and catalogs from over 120 countries.
- *Google Analytics:* Free log and e-commerce analyzer for mobile sites that can provide rich insights into traffic and marketing effectiveness. The service can track mobiles visitors, as well as sales and conversions; it can measure site engagement goals against threshold levels defined by webmasters, developers, marketers, and brands. The analytics service can trace transactions to campaigns and keywords; it can generate loyalty and latency metrics.

  The service can compare site usage metrics with industry averages; it can track Flash, video, and social networking for mobile sites. Google Analytics supports advance segmentation, flexible customization, and e-commerce tracking. The mobile site analytics can uncover trends, patterns, and key comparisons with funnel visualization and motion charts; this analytical tool complements a suite of Google-related products, such as their AdNet, DoubleClick, and Google+ networks.
- *Nihuo:* Shows the "who, what, when, where, and how" of visitor behaviors for mobile sites. This log analyzer is fast and powerful for small- and medium-size mobile sites. It can report on where website visitors come from, which pages are most popular, and which search engine phrases brought visitors to the site. The software can analyze logs generated by Apache, IIS, Ngnix, and lighttpd Web servers.
- *SAS:* Web analytics and reports from the world's largest statistical company. Their mobile site analyzer is a scalable server solution for business reports and data interactive visual reports for iPhone, iPad, and Blackberry Torch or Bold devices; it does not support Android mobiles.
- *Webtrends:* Measures anything happening on a mobile site. It can import data from Facebook and Twitter, and can compare apps and mobile sites in

support of all major mobiles. Webtrends provides the ranking of mobile site pages and enhanced measurement for apps; it can measure Facebook data stored for long-term trending and iTunes App Store downloads and revenue reporting. The software can monitor millions of blogs, tweets, video-sharing sites, Facebook, and other mobile site activities to find conversations pertinent to brands (Figure 5.24).

■ *SeeVolution*: A heat map technology mobile site analytics firm. Click heatmaps are used to analyze how people click in a mobile site; they radiate the section of a site that is most hot and active. This software works with Google Analytics to generate visually reporting heatmaps of mobile visitor activity.

■ *ClickTale*: Also provides heatmaps to break down mobile interactions and behaviors. The heatmaps capture every mouse move, click, scroll, and keystroke, using a tiny piece of JavaScript copied into a mobile site. The whole process is completely transparent to the end user and has no noticeable effect on site performance. There are also log analyzing services such as Clickdensity and Crazy Egg that provide hosted heatmap solutions.

Figure 5.24 Webtrends reports on multiple types of mobile transactions.

5.20 Social Mobile Consultancies

Engagement mobile marketing uses influence metrics to measure a consumer's likelihood to encourage friends to consider, recommend, or purchase a brand via their mobile. In a social mobile world, conversations about brands or companies provide a new metric for marketers and enterprises. Mining mobiles can result in the development of consumer profiles of brand champions. The following boutiques can assist mobile miners and marketers in leveraging these new social interactions by providing vital strategies, campaigns, and metrics.

- *TNS Global:* A market research company that provides customized reports organized around their clients' specific industry sector, including mobile activities in over eighty countries. In each country, they combine the benefits of industry specialization and research expertise to deliver marketing insights. Their technology team consists of over 500 researchers across 60+ countries. TNS Global helps clients define brand strategy by identifying category dynamics, brand positioning, and consumer needs to make their brand stand out from the competition.
- *MotiveQuest* (<http://www.motivequest.com/>): Monitors and reports on "brand buzz" to understand customer motivations (Figure 5.25). They concentrate on understanding prospects and customers so that clients can grow their market share. They call their social research approach "online

Figure 5.25 A MotiveQuest map of brand motivations.

anthropology." With their MotiveQuest Framework, clients can discover what motivates groups of mobiles and what they can do to turn them into brand and company advocates.

■ *Attensity*: A software company and marketing consultancy specializing in analysis and reporting on social messages using advanced text mining algorithms. Leading brands recognize that their customers are both passionate and vocal about their products and services, and they are sharing their experiences in mobile conversations every day on social networks. Attensity provides the text mining technology to capture unstructured digital conversations taking place via mobile sites and apps (Figure 5.26).

Mobile data miners and brands need to create profiles of product and services "champions" in order to target, recruit, and retain more of these valuable "evangelists" with similar interests and preferences using clustering and classification software. Mobile marketers, brands, and enterprises need to focus on identifying these "brand zealots" via mining mobiles in order to attract more with similar preferences, desires, demographics, and influences.

Mobile marketers and developers need to look at evangelist consumers as co-marketers for their brand or company. Through the strategic use of engagement marketing, companies and brands can invite consumers to become the co-producers of marketing campaigns, such as the creation of slogans, designs, apps, tweets, widgets, videos, and other types of viral communications via their mobiles. They need to encourage and reward them when they invite others via social networks.

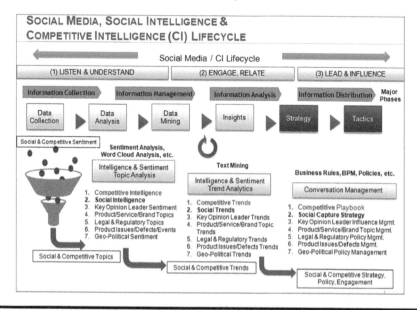

Figure 5.26   The Attensity process.

5.21 Mobile Analytics

The future of mobile analytics will gradually shift from simply modeling the behaviors of millions of mobiles in people's pockets and purses to those involving how they interact with their devices in ways that do not involve tapping keys and swiping a screen. Mobiles are broadcasting beacons about how, when, where, and what humans want. The average American now spends as much or more time with their mobile device than with any other form of media, bar television.

According to a report published by the mobile analytics firm Flurry (Flurry.com), while American consumers spend 40 percent of their media time watching television, mobile is now in second place, commanding 23 percent of their time. Despite this huge increase in engagement with mobile devices, brands, companies, and marketers have only committed approximately 1 percent of their spending to mobile campaigns. Comparing where usage and spending vary, most is in over-spending on print advertising and—even more severe—under-spending in mobile, the report concluded. The importance of this is that the data mining of mobile devices is also under-developed or nonexistent.

Unlike print advertising, mobile marketing can be precisely calibrated to reach and persuade consumers by the data mining of their mobile devices. In short, despite the fact that mobile advertising is growing, the platform is far from getting rational levels of spending compared to other media. There is really no modeling of mobile big data taking place aside from mobile site reports and the counting of app downloads, although several companies are developing new mobile analytic strategies and tools. With the Apple and Android app economy only a few years old, Madison Avenue and brands have yet to adjust to an unprecedented adoption of apps by consumers, and the same can be said of mobile sites. More than 1.2 billion apps were downloaded across Android and Apple devices worldwide, according to Flurry (Flurry.com), the highest number ever recorded by the firm (Macys.com) during the Christmas season of 2011.

Gradually, mobiles will become more intelligent—smart enough to integrate simple commands and, via their triangulation capabilities, to respond and remember simple dictation, such as "Remind me to call Mom when I get home." The potential for this trend is already here with Apple's AI-based Siri and its ability to understand not only natural language input, but also context. The mobile will keep track of the time, place, and patterns of prior calls before reminding its human owner to give Mom a ring.

While voice assistants such as Siri may seem revolutionary today, they are actually rather simplistic compared to what is available with AI and new multimodal and hyperbolic capabilities of mobile interactions. Mobiles are already full of sensors that input interest and location information; the next frontier will be their ability to use their cameras to find a bus stop, or a restaurant—and have a bus schedule or menu instantly delivered.

The mobile accelerometers in these devices can be trained to understand and learn gesture-based commands, so that as the user is about to go leave his house, like a Jack Russell Terrier, mobiles can learn that a flick forward means "let's go for a ride and start the truck," or that a flick backward means "activate the house alarm."

Mobiles' internal sensors such as gyros, GPS, accelerometers, Wi-Fi, and external geo-fences and other triangulation techniques, coupled with AI can enable mobiles to *learn* their owner's behaviors, thereby enabling them to anticipate and advise them about personal options in their lives and becoming their *personal assistants* to respond to "Where can I go for lunch without gaining any weight tomorrow? I have $3.50 to work with, and I only have twenty minutes." Multiple databases would be dinged for location, health, coupons, and cheap grub information delivered in microseconds to that mobile.

Not all digital data has the same value. The average person is bombarded with marketing data during almost every waking moment: TV ads, text messages, radio broadcasts, various print media, and a whirlwind of instant messages, pop-up ads, videos, and sound-bytes on the Web. No one can accurately process or make use of all of that information—that is over 5 trillion megabytes, and that is where mobile analytics data-driven apps and mobile sites automating decision making comes in.

Highly advanced algorithms, like SOM and CART, process both structured data, which allows developers to select specific pieces of information based on columns and rows in a field organized and searchable by data type within the actual content, and unstructured data that is not part of a database to identify and analyze patterns that lead to targeted insights for supporting decisions. These algorithms allow brands and developers to take a "big-picture" perspective of the increasingly growing mobile data, which we in turn can organize and process in a way that the human mind could not hope to comprehend by itself.

Aristotle theorized that government was the greatest human endeavor because it influences and controls well; that same logic can be applied to the implications of mobile analytics, which has emerged as a universally applicable method for aiding and making all manner of decisions. Analytics can assist the decision-making process in virtually any field, discipline, or industry, making it one of the most commonly used and relied-upon methods to inform wide-ranging leadership. Because of this, it is not unrealistic to imagine that analytics can become the primary method for all major world organizations to craft their policies and plans of action.

The implications for knowledge-based professions, government, and consumer-driven companies will be extraordinary. Never have the following mantras been more relevant: He who knows his customer best has no competition. Knowledge without relationships has no value. The power of analytics enables corporations, government, and the brands to align their products, services, and content with the expectations of the consumer. This is truly evolutionary because mobiles not only dictate what they want, but where and when they want it.

Mobile analytics is being applied to a vast number of areas of knowledge: business, ecology, healthcare, sports, government policy, city planning, technological development, medical research, and practically every scholarly discipline imaginable. Mobile analytic systems are already being used all over the world to increase the efficiency and success of some of the most important endeavors on the planet: air traffic control, disease control, water and utility operations, evacuation planning, food and utility distribution, auto traffic planning, medical diagnoses, banking, mortgage processing—and now marketing and advertising.

In theory, data mining of mobile devices technology is only as good as the people who create and program it; creative use of clustering, text, and classification software is required. Fortunately, the mobile analytics field has attracted some of the most brilliant technological minds in the world to ensure accuracy and relevance to consumers. Reliable studies have shown that brands and companies that use advanced mobile analytics as integral parts of their decision making greatly increase their productivity and their profits. Other research has yielded similar results, indicating that the accuracy of mobile analytics applications is considerable.

Of course, nothing is infallible, especially when it comes to predicting the future. Predictive mobile analytics, however, have shown a high degree of accuracy. Generally speaking, mobile analytics have proven to be highly accurate, including important, practical applications that depend on high degrees of precision such as air traffic controlling, utilities, medical applications, etc. All these applications have resulted in spectacular success. The future, however, is in mobile marketing and advertising, which will be a multi-billion-dollar sector.

But the variety of apps of mobile data mining technology is only one aspect of the overall versatility of analytics. It is also versatile in terms of its use when considering *cloud* analytics. By utilizing cloud-based analytics systems, users have more options, and more protection, than ever before. Users can have full access to all services from any device anywhere on the planet. They are also free from many of the problems that plague users of conventional computer systems: limited memory, limited storage, loss, theft, accidental damage, viruses, and various other complications and catastrophes.

As companies adopt cloud-based wireless systems, the world's global economy will transition into a sleeker, faster, more mobile, and more flexible work environment for all industries. Massive amounts of information will be analyzed through cloud systems, thus enabling developers and miners to tackle more complex, wide-ranging problems and questions than ever before. The unlimited versatility of such mobile mining systems indicates a quantum leap beyond merely plugging in and surfing the Internet. Technology of this level is a prelude to a world in which advanced research and thinking are possible on levels that have never been imagined.

While the "information age" gave people greater access to raw data than ever before in history, the "analytics age" goes beyond the mere presentation of jumbled data and provides something that is far more valuable: insight and practical applications for marketers and brands. Everyone, from policy makers and business owners to private citizens, may soon be able to log on to any server via their mobile device from anywhere and run sophisticated, customized analytics programs to gain new perspectives on a number of issues pertinent to their life and work.

Analytics does something amazing. It converts titanic amounts of raw data into accurate, useful, applicable insights, and represents the next evolution of the information age. Daily, we create with every keystroke 2.5 quintillion bytes of data from sensors, mobile devices, online transactions, and social networks. Monthly, we express ourselves through 1 billion tweets and 30 billion pieces of content on Facebook. Some call this "big data," but astute brands and enterprises realize that this is a unique market opportunity.

The advent of technologies such as software and hardware that instantly analyze natural human language, and cluster and segment massive amounts and varieties of big data flowing from sensors, mobile devices, and the Web, are helping today's data pioneers find answers to questions such as, "How do consumers feel about my product?" or "Why are mobile Web visitors not staying?" or "What happens when consumers enter a geo-fence?" These systems literally sift through the data and identify patterns and trends on-the-fly, then present them in a way that is easy for people to understand.

Mobile trends can then be fed back into systems for further analysis that allows for new kinds of questions to be asked, such as "What will consumer reaction be if we introduce these kinds of products?" or "How will app downloads help in these emerging markets?" or "Why does sustainability positively impact our mobile business model over the next five years?"

With big data comes the new role of the data scientist. But forget about an image of a scientist in a white labcoat; data scientists are our modern-day data miners and app and mobile site developers in the business world. They are individuals or teams at companies who sift through all the data coming in from everywhere with the goal of gaining insights into consumer sentiment and other tough business challenges, which in turn can provide a competitive advantage for their brand and company.

Data scientists see mobile trends and can calibrate how consumers buy things and what actions can make a business or brand successful. Today, mobile information is gold. Knowing purchase preferences and intent to buy can help organizations and brands make smarter business decisions, and keep their customers loyal to their brands and offer them the right promotions via their apps and sites.

One of the greatest challenges facing data miners is determining the most important information to look at; this can be accomplished by the strategic use of SOMs and decision trees. In the past, companies and brands looked in the "rearview mirror," collected information from social media sites, and stored it inside a database. Then they analyzed it, which could take weeks, and brought those insights back into the business and company. However, as streaming analytical software has shown, that can be accelerated and take place in real-time as events take place.

Now that businesses can analyze any information as it happens, they can stop looking in the rearview mirror and focus on the road ahead, which in the mobile world is immediate. We are at a unique point in time where companies and brands can better understand their customers as they share their feelings about a product, a service, content, and their customer experience.

The role of data scientist is causing shifts inside organizations and across business cultures, making the job in great demand. Currently, there are 10,000 job openings from a broad variety of companies, ranging from deal-of-the-day websites, to traditional retailers, to global consumer goods distributors—but most importantly, the ever-expanding mobile world.

Mobile data miners must play a central role not only in handling mobile information complexity, but they must also coordinate their findings with both business, mobile, and IT groups. While it is good to have some background in math, modeling, and analysis, successful data scientists can work across the organization and influence various business and marketing entities. It is this ability to work in technology and push collaboration that will help mobile data miners succeed.

We have heard about analytics before. Now with advancements in big data and mobile technologies and the emerging role of data scientist, companies do not have an excuse when it comes to improving customer service or delivering new products and services that better meet consumer needs in the mobile world. After all, this kind of analysis will generate even more consumer sentiment and brand loyalty, which can then be analyzed and acted upon for more revenue growth.

Big-data technologies and new roles such as the "data scientist" will enable companies and their employees to embrace mobile analytics for business growth. We live in the era of information and the trends that are hidden in the streams of data points. Those who ask the right questions and apply the right technologies and talent are certain to crack the curious case of big mobile data. Mobile is instant and not very tolerant of data latency, so that microsecond marketing and precise targeting is the standard.

Mobile analytics will move ahead as more and more devices, appliances, homes, autos, TVs, and other gadgets gain mobile connectivity, leading to the creation of "swarming intelligence" techniques and strategies, by which data mining is used to replicate the way biological systems communicate and organize themselves. Not only will mobiles be instruments of communications, they will become beacons of what their human owners want, when and where and how they want it; they will be intelligent enough to machine learn their human owner's psyche and become incredibly powerful diaries of their personal life.

Mobile marketing revenues will grow to $58 billion by 2014 (Gartner.com), there will be more mobile Internet users in 2015 than PC users (IDC.com), mobile marketing will grow by 30 percent in the next 5 years to $1,047 billion in 2016 (Ovum.com), mobile traffic will increase tenfold over the next 5 years (Ericsson.com), and Cisco (Cisco.com) forecasts (via its Visual Networking Index) that traffic alone will grow to fifty times more than it is today by 2016 and there will be more mobiles (10 billion) than the number of people on Earth (estimated at 7.3 billion)!

The data mining of mobile devices has a great future, so make the most of it. Good luck!

Index